Medizinische Länderkunde
Geomedical Monograph Series

4
KUWAIT

Springer-Verlag Berlin Heidelberg GmbH 1971

Medizinische Länderkunde

Beiträge zur geographischen Medizin

Geomedical Monograph Series

Regional Studies in Geographical Medicine

Schriftenreihe der / Series of Monographs of the
Heidelberger Akademie der Wissenschaften · Mathematisch-naturwissenschaftliche Klasse
Begründet von / Founded by

Ernst Rodenwaldt †

Herausgegeben von / Edited by

Helmut J. Jusatz

Professor Dr. med., Direktor des Instituts
für Tropenhygiene und öffentliches Gesundheitswesen am
Südasien-Institut der Universität Heidelberg

Unter Mitarbeit von / In collaboration with
Dr. phil. BERTHOLD CARLBERG, wissenschaftl. Kartograph, Murnau/Obb. · Dr. rer. nat.
HEINZ FELTEN, Säugetierabteilung des Forschungsinstituts Senckenberg, Frankfurt/Main ·
Prof. em. Dr. med. LUDOLPH FISCHER, Direktor des Tropenmedizinischen Instituts der
Universität Tübingen · Prof. Dr. phil. HERMANN FLOHN, Direktor des Meteorologischen
Instituts der Universität Bonn · Prof. Dr. phil. GERHARD PIEKARSKI, Direktor des Insti-
tuts für medizinische Parasitologie der Universität Bonn · Prof. Dr. rer. nat. ULRICH
SCHWEINFURTH, Direktor des Instituts für Geographie am Südasien-Institut der Univer-
sität Heidelberg · Prof. em. Dr. phil. Drs. h. c. CARL TROLL, Direktor des Geographischen Instituts der Universität Bonn

KUWAIT

Urban and Medical Ecology

A Geomedical Study

By

Geoffrey E. Ffrench and Allan G. Hill

M. A., M. D. (Cantab)   B. A., Ph. D. (Dunelm)

With 61 Figures, 26 Text Figures,
56 Tables, and 3 Maps

Springer-Verlag Berlin Heidelberg GmbH 1971

GEOFFREY E. FFRENCH, M.A., M.D. (Cantab), Fellow of the Royal College of Physicians of Canada, Fellow of the American College of Physicians, Fellow of the Royal Geographical Society, Consultant in Industrial Medicine, Central Middlesex Hospital, London, formerly Chief Medical Officer, Kuwait Oil Company, Kuwait

ALLAN G. HILL, B. A., PH. D. (Dunelm), Lecturer in Geography at the University of Aberdeen

Additional material to this book can be downloaded from http://extras.springer.com.

ISBN 978-3-642-65174-8   ISBN 978-3-642-65172-4 (eBook)
DOI 10.1007/978-3-642-65172-4

Herstellung der Karten 1—3 in der Geomedizinischen Forschungsstelle der Heidelberger Akademie der Wissenschaften. Druck im Kartographischen Atelier von Henning Wocke, Karlsruhe.

Preface

The developing countries are receiving generous attention from experts, officials and academics drawn from a wide spectrum of specialist interests. Some of this effort is directed towards a solution of several of the world's most pressing problems, including ill-health, under-nourishment, and rapid population growth, but other workers are more concerned with the less immediate but nonetheless very significant theoretical aspects of the developing countries. This book is an attempt to bridge the gap between these two approaches.

At this present juncture in time we are faced with the realization that the experience of Europe or North America may be of limited assistance in the interpretation of current trends in the developing world. Not only is the sequence of events novel but even the events themselves may be occurring for the first time in history. This is apparently the case with urbanization and is one of the reasons for this present study. Before any new theory can be conceived, we need as much information about the dominant processes at work throughout the developing world. Kuwait's unique experience requires careful analysis if we are to understand the significance of parallel situations elsewhere.

There is another motive for a study such as this which is the conviction that in the understanding of many of our contemporary problems, an inter-disciplinary approach is required. In this instance, the process of urbanization clearly has implications of a slightly different character for specialists in any single field. Here it is hoped that the skills of the physician can be usefully blended with those of the geographer to highlight some of the advantages of a holistic approach. Perhaps others will be encouraged to enter this exhilarating and rapidly expanding field. Previous to this book no other approach in depth had been made along these lines, not withstanding the informative though restricted account of Kuwait in Simmons, Whayne, Anderson and Horack's "Global Epidemiology" published over sixteen years ago.

Our first debt is to the Government and people of Kuwait who co-operated with us so willingly in their several capacities to make this study possible. Many officials and individuals in a wide range of Ministries,

Government Offices, and commercial organizations deserve our sincere thanks for their attention to our many enquiries. In particular, we would like to thank the officials of the Planning Board and the Central Statistical Office, Kuwait Municipality, University of Kuwait, and the Kuwait Oil Company. The following individuals deserve our special thanks: Mr. Ahmad al-Duaij, Mr. Fouad al Hussaini, Mr. Hamid Shwaib, Mr. Abdulaziz al-Hamdan, Mr. Fouad Haddad, Mr. Ahmad al-Haj, Mr. Marwan 'Adra', Mr. Muhammad Sukhon, Professor Abdul Fattah Ismail, Professor Dawlat Sadiq, Professor Muhammad Mutwalli, Dr. Muhammad Sharnubi, His Excellency Ibrahim Shatti, Dr. Noel Brehony, Professor W. B. Fisher, Dr. John Brebner, Dr. Alan Horan, Mrs. Sheila Robinson, Miss Maggie Austin, Dr. A. D. Ezzat, Dr. Eli Chalhoub and Dr. Y. Shaker.

We are particularly grateful for the permission given to us by Mrs. John Van Ess to quote from the unpublished memoirs of the late Dr. Stanley Mylrea.

Without the continued support, encouragement and help with the preparation of the manuscript from our wives, we could not have achieved our aim. Thank you.

We cannot close without mentioning the late Dr. Saba George Shiber whose vigour and interest in every facet of urban growth in Kuwait and in the whole Middle Eastern area, did much to spark off this study.

We are grateful to the Carnegie Trust for the Universities of Scotland and to the University of Aberdeen for financial assistance towards the cost of map production.

The publication of this work has been made possible by the courtesy of Professor Dr. med. Helmut J. Jusatz, the editor of the Geomedical Monograph Series — Regional Studies in Geographical Medicine. Our thanks are due to him and his team of the Geomedical Unit of the Heidelberg Academy of Sciences for their support in the editorial work and to the publisher Springer-Verlag Berlin-Heidelberg-New York, for the layout of this book.

GEOFFREY E. FFRENCH
ALLAN G. HILL

Contents

Tables

Maps

Introduction

Over the past twenty-five years the oil-producing countries of the Middle East have become of vast importance to the development not only of Europe and Africa but also of the Far East, particularly Japan. Their political and cultural development over this short space has been of a similar or even greater magnitude to that which the countries of Europe under-went over a much more extended period of time. Throughout the Middle East radical changes have occurred in the composition and distribution of population and society in the last few decades—changes which have a significance extending far beyond the regional and cultural context in which they are rooted.

In the Persian Gulf * region however, the pace and scale of change has been greater than elsewhere in the Middle East because of the discovery of huge oil reserves in the post-1945 period. While British interests have been represented in the area since the signing of the Exclusive Treaties in the mid-nineteenth century, the oil reservoirs acted like magnets in attracting to the Gulf states European and North American funds, personnel, and equipment in unprecedented quantities. The juxtaposition of Western capitalism and Middle Eastern feudalism produced some stark contrasts. Although culture contact is not a new phenomenon, in the Gulf States the convergence of cultures was particularly marked because of the period of contact and because of the medium through which Western concepts were introduced into Arabia. Few areas of the world can have been as isolated from European influence and colonization as the Gulf States were in the early twentieth century and there can be no more "modern" economic activity than the international oil industry. As a result of these circumstances, a sequence of changes began in a variety of apparently unconnected spheres which we are now in a position to identify as a cohesive pattern. With rising national and personal wealth, new demands were created, new patterns of life emerged, while the values of the traditional society were seriously challenged. The sequence of change is akin to a series of innovation waves which gather momentum in the northern Gulf and then sweep southwards through the Trucial States to the Sultanate of Muscat and Oman.

In a sense, Kuwait is the archetype for the whole Gulf region. This country in the last three decades has charted the course of development for its smaller neighbours in the lower Gulf. In particular, the impact of concepts and commodities derived from the West has been felt first in Kuwait and subsequently in the other Gulf States. The examination of Kuwait as a leading edge in the modernization process would be justified on these grounds alone but there are a variety of additional aims in this study.

Foremost is the task of indicating, through the spectrum of medical geography, how in the study of such processes as modernization and economic development, the barriers between the established academic disciplines must be taken down.

Progress in apparently unconnected fields has achieved changes in the pattern of life—its longevity and infant mortality—and in the general physical development of the people of these countries, which together with a massive immigration from less favoured areas, has produced an enormous problem for administration and long-term planning.

The complexity of the process of change is bewildering because of its many dimensions. Several of these dimensions have already been dealt with by other authors.

Urbanization and economic development are among the most popular topics dealt with in the contemporary literature of the social sciences. A growing awareness of the complimentary nature of the developed and developing worlds has sparked off numerous specialist studies investigating the conditions and causes of differential levels of national prosperity; in fact what decides a favoured or less favoured nation is not invariably the possession of valuable raw materials. Since however, an increase in prosperity seems almost inextricably linked with industrialization and ultimately urban growth, the evolution of urban centres is a necessary corollary of many of these studies of economic development. Further, the accelerating pace of world urbanization, which by mid-century had resulted in the concentration of twenty-one per cent of the world population in towns containing over twenty thousand inhabitants, has attracted the attention of students eager to describe and interpret this world-wide phenomenon. Such a close scrutiny of the developing world has clearly had two major repercussions.

First, as detailed information on specific topics and regions becomes available in increasing amounts, it is clear that many of the laws and generalizations evolved in the developed world are inapplicable in the context of the still developing countries. Second, despite attempts to synchronize phases of development—for example, by equating political, economic and demographic conditions in nineteenth century Europe with those prevailing in developing countries today—comparison shows that more than mere historical time is responsible for the present differences between the contemporary worlds.

An important stage in the study of the links between both these worlds has been reached, which indicates

---

*) In this book the term "Persian Gulf" is used to refer to the inland sea separating Iran from Arabia. It should be pointed out that among the inhabitants of the Arabian shore and in the Arab world as a whole the term "Arabian Gulf" (Khalij al-Arabi) is in common use. The use of "Persian Gulf" here however, conforms to accepted usage elsewhere and implies no political bias.

clearly that the developing world should be dealt with as an independent unit rather than a deviant of the conventional European and North American models. Many of the processes affecting the poorer countries today can be justly dubbed unique; few direct parallels with the richer countries are possible but those that are, can sometimes be used to advantage, particularly in coping with the problems of rapid population growth and the maintenance of high standards of public hygiene despite sizeable rural-urban migration.

Kuwait has been the subject of many articles and several books. Most of the emphasis has so far been placed on the country's wealth and the speed with which the traditional society has been swept away by the economic effects of the modern and mammoth oil industry. These two themes are obviously inescapable in any consideration of Kuwait, but they emphasise Kuwait's exceptional situation while missing many general points of at least equal significance. Doubtless Kuwait has been an anomalous development, a small state, apparently barren, yet now with one of the highest per capita incomes in the world. But in recent years, Kuwait's place in the overall development of the Middle East area has been accepted while yet smaller sheikhdoms have been discovered with apparently much greater potential wealth. Rapid economic expansion is in fact becoming a less exceptional phenomenon as we can see in Japan, Hong-Kong, Brazil, Venezuela, Libya and Alaska, although most of the developing world still displays extremely low rates of growth of the per capita gross national product.

Instead of pointing to the few "boom" states of this developing world as exceptions with little general significance, it is suggested that more studies in depth are required to bring out not only these exceptional factors but also the more general lessons of value to others. Kuwait provides a stringent testing ground for many theories, and we propose to examine some of these critically.

I. Urbanization and Population Growth in the Middle East

1. Definitions

Urbanization is as old as civilization and equally complex. It is a continuing process with world wide manifestations, making comprehensive accounts of its causal factors and geographic occurrence as elusive as its factual definition. But urbanization is not a uniform process operating through time; with a time span extending from the Neolithic to the present day, and a geographic spread covering the entire occupied surface of the earth, almost any generalizations will be severely stretched to encompass even the salient aspects of such a process.

Nevertheless, some general statements on the nature and causes of urbanization are generally accepted by archaeologists, economic historians, geographers, and contemporary sociologists alike.

In both rich and poor countries, two processes radically affecting population and society are distinguishable; urbanization and urban growth. Urbanization as a process is defined as an increase in the proportion of a nation's population located in urban areas [1, 2]—while urban growth implies an increase in the population of towns although the balance between urban and rural dwellers may remain largely unaltered throughout [3]. Of the two, urbanization is more common and involves greater structural re-organisation, since as well as the geographical re-location of population it also implies an adjustment in the employment structure of a nation. This aspect of urbanization provides the theoretical rationale for linking such processes as economic change and development with urbanization, for it is argued, if fewer people are employed in agriculture, more will be employed in manufacturing and services [4]. Such a sectoral re-deployment of labour and hence capital is usually regarded as forming the basis for economic modernization in a national economy. Observations confirm this assumption, for "rich" countries have relatively few agriculturists while the reverse is true for "poor" countries.

While such a generalization, by equating poverty with agriculture and wealth with industry and services seriously misrepresents agriculture as a whole and several countries in detail (e. g. Denmark), it stresses the most important point about urbanization which is, that as urbanization involves not only population movement but also a sectoral re-deployment of the employed population, urbanization inevitably has *prima facie* associations with economic development.

Historically these connections can be readily confirmed. The division of cities into "pre-industrial" and "industrial" formalized the distinction between the two phases of world urbanization associated respectively with the Neolithic Revolution and the Industrial Revolution. The factors involved in this close association between economy and urban life during the Neolithic period have been summarized by Hawkes & Wooley.

"It is an axiom of economic history that real civilization can only begin in regions where the character of soil and climate makes surplus production possible and easy; only so is man relieved from the necessity of devoting all his energies and all his thought to the problem of mere survival, and only so is he enabled to procure from others by means of barter those things which minister to well-being and promote advance but are not naturally available in his own land; moreover, such conditions must prevail over an area large enough to maintain not merely a small group of individuals but a population sufficiently numerous to encourage occupational specialization and social development. So does civilization begin. Most of the community continue to devote their energies to actual food-production, but those whose gifts or tastes are of another sort become artisans, specialists in production of a different but scarcely less necessary kind, making those things without which the agricultural worker cannot get on" [5].

In the second major phase of world urbanization, occurring in the later eighteenth and nineteenth centuries in the West, new urban centres grew up alongside older but expanded towns, the growth of which was precipitated by technological innovations—particularly the use of steam power—introduced by the Industrial Revolution. Improvements in communication made possible the organization and administration of unprecedently large urban agglomerations. Today this process of agglomeration continues, resulting in some cases in what has been called "Megalopolis" [6].

However, our belief is that these established relationships affecting economic development must be questioned in several instances in the developing world.

While most forms of economic development require some geographic re-location of the labour force, the inverse assumption sometimes made—that a drift of people from rural to urban areas is associated with economic growth—is in some cases erroneous. Azeez [7] has shown in Iraq that just this form of movement (from the countryside to Baghdad) can act as a negative rather than a positive stimulus to economic development.

We believe that there is a need to distinguish two processes often confused as one. These processes are first, true *economic growth* bringing a substantial per capita increase in real wealth to the total population; and second, *economic change* bringing only dislocation of the pre-existing economic system usually as a result of some form of contact with the developed world. Between the two there is no sharp dividing line, for in every world economy there is economic growth of a sort, comprising factors operating in alternating directions. Thus the crucial question is to decide whether the present economic shift in a nation's economy is a step likely to bring sustained per capita benefits as well as other material and cultural rewards. Such a holistic approach to economic development is upheld by most economists because of the indivisibility of economic and total welfare. In drawing this distinction between economic growth and economic change, it is imperative to resist the tendency displayed in many studies where development for the poorer nations consists of imitating their more prosperous peers. Interest must be centred on the most efficient combination of land, labour and capital without introducing the bias in the viewpoint of the Western world.

The assumption that urbanization can take only one form must be criticized. Based on economic history derived largely from Britain in the nineteenth century, the associations between mechanization, factorization, industrialization, and urbanization are assumed to be causal. Less clumsily, an urban area of industrial and related developments is seen as a region of major labour deficiencies, which are met from the surrounding rural areas. Such a process of rural-urban migration leads to population concentration—in short, urbanization.

However, we know from several case studies of rural-urban migration that a multitude of factors are involved [8]. It is suggested that just as economic development and economic change were distinguished above, urbanization and another process called "pseudo-urbanization" [9] should also be distinguished. By urbanization is implied an increase in the proportion of a population in urban areas, taking natural increase and migration into account, but implying also that the process reflects a true economic need for population relocation enabling

a country as a whole to progress to a more evolved level of economic development. Again it seems we are in danger of generalizing from Western-derived models but we must recognize the two processes operating throughout the world today. For by "pseudo-urbanization" which is taking place in parts of the underdeveloped world in varying degrees today, we imply that the geographical re-location of population takes place but without the concomitant changes in the economy which are elsewhere part and parcel of the process. Rural dwellers today may move city-wards with no intentions of joining the industrial society of which geographically they have become a part. Migrants are often ill-equipped to perform even the simplest manufacturing tasks although the opportunities for employment are available in the first place. Hence in the urban areas, they subsist by performing personal services and indulging in minor retailing, living off the industrial society as parasites instead of being absorbed by it and its occupational structure. For this reason, employment in this service sector is no guide to the evolution or sophistication of an economy, as Kuwait demonstrates.

In the "poor" countries whose development process is only beginning, changes are occurring which differ from the model system described for Europe above. First, natural increase of population is a sizeable factor in virtually every developing nation. Since the advent of scientific medicine preceded the Industrial Revolution, both of which were almost contemporaneous in Europe, population growth rates of over 2.5 per cent per annum are common. Second, feeding these rapidly expanding populations poses real problems because no Agricultural Revolution preceded this phase of demographic growth. Third, even though the techniques and machinery for progress are available, the levels of literacy and general advancement of the populations of the developing countries renders these aids unavailable to them. As a result, manufacturing plays only a small part in the total economy, fulfilling the requirements of one section of the population rather than acting as a catalyst to developments in other sectors. Economists have recognised this evolution in many of the countries of the developing world, calling it "dual development" [10] leading to the formation of "dual societies" [11]. Such an evolution obviously warrants close attention because of both its practical and its theoretical significance.

It appears that our notions about economic development are coloured extensively by the experience of the nations of Western Europe in making the transition from an agricultural and handicraft society to the fully fledged industrialized and urbanized society today. Throughout the eighteenth and nineteenth centuries, with only a moderate increase in total population, the factories of Europe had growing demands for labour. The prime source of this labour was the surrounding rural areas. Other factors were involved in this rural-urban migration but the differences between wages is the prime motivation in such a pattern of migration. Hence, manufacturing and service industries grew at the expense of agriculture which maintained or increased its productivity by application of factory-derived machines instead of manual labour to production methods. In all, the reciprocity implicit in this classic chapter in world economic history produced a harmonious blend of urban and rural life which apparently contributed largely to Europe's ensuing success and prosperity.

2. Middle East Urbanization: Extent and Special Characteristics

To confirm or refute the ideas presented above, many studies in depth of the urbanization process and its correlates are obviously called for. Such a wide ranging task is beyond the scope of the present work, but before embarking on a close analysis of the Kuwait urbanization, it is of paramount significance to set this study in both its regional and world-wide contexts. In order to compare levels of urbanization internationally, a heavy reliance must be placed on the urban population statistics provided by the United Nations, since they are the only directly comparable figures available.

As an alternative, we can use the rural population and subtract it from a country's total population to give us a figure which not only incorporates all centres of 100,000 and over, but also smaller centres which are important constituents of that country's urbanized population. This, we believe, produces a more sensible ordering of individual nations by degree of urbanization than other methods using size thresholds.

Table 1 shows how the world urban population has increased very rapidly in the last 150 years.

Table 1. *Total world population and world urban population 1800—1960*

Year	World population (millions)	Population in cities of 5,000 and over (millions)	Percent in cities over 5,000	Percent in cities over 100,000
1800	906	27.2	3.0	1.7
1850	1171	74.9	6.4	2.3
1900	1608	218.7	13.6	5.5
1950	2400	716.7	29.8	13.1
1960	2995	948.4	31.6	20.1

Sources: 1. Davis, K., Hertz, H.: U.N. Report on the World Social Situation 1957, p. 114.
2. U.N. Demographic Yearbook 1962, Tables 9 and 10.

Of greater immediate significance is the regional variation in the proportion of the population in urban places. The definition of an urban place varies widely—some countries basing their definitions on administrative boundaries and the status of the area they enclose and others treating any area with a population agglomeration as low as 250 (e. g. Denmark) as an urban place. Unfortunately, the United Nations only collect internationally comparable statistics for urban centres which contain over 100,000 people. This arbitrary threshold—particularly in the Middle East—produces anomalies. For example, Tripoli in Lebanon recently surpassed 100,000 in population with the result that Lebanon's index of urbanization leapt from 16.3 to 29.8 per cent from the mid-1950's to the mid-1960's using this method of definition for urban areas.

A recently published monograph [14] allows us to bring together information on urban populations first for the world and secondly, for individual nations of the Middle East. Overall, it seems from Table 2 that Northern Africa and Southwest Asia (the Middle East) are less urbanized than two-thirds of the world's major regions. However, these average figures for Northern

Table 2. *Two measures of world urbanization in 1960*

World region	Population in urban areas (per cent)	Population in cities of over 100,000 (per cent)
Australia — New Zealand	77.7	53.4
Northern Europe	72.2	57.3
Northern America	69.7	49.8
Western Europe	68.0	41.0
Temperate South America	65.0	45.5
Japan	63.5	41.9
U.S.S.R.	50.1	24.9
Eastern Europe	48.4	20.5
Middle America	46.2	17.3
Southern Europe	45.5	24.1
South Africa	44.8	26.5
Tropical South America	44.7	23.3
Caribbean	38.4	17.3
Northern Africa	29.6	18.2
Southwest Asia	29.5	15.1
East Asia	18.0	11.8
Southeast Asia	16.6	9.7
South Central Asia	16.4	8.5
Western Africa	14.7	5.0
Middle and Southern Africa	11.6	4.5
Eastern Africa	7.5	2.8
Oceania	6.0	—

Note: The definition of the urban areas is usually that of the countries concerned.
Calculated from: Davis, K. [13].

Africa and Southwest Asia conceal a great variety of levels of urbanization within these two regions (Table 2).

3. Levels of Urbanization in the Middle East

With levels of urbanization (percentage in urban areas) ranging from some of the highest values in the world (over 70 per cent) to values below those of Eastern Africa and Oceania, the world's most "rural" regions, it seems we can divide Middle Eastern nations into four major groups on the basis of Table 3.

Group A: Over 40 per cent in Urban Areas

This group consists of five small countries (all under 83,600 sq. km in area), four of which are now oil-producing states in the Persian Gulf. These states (Bahrain, Kuwait, Qatar, and the Trucial States) have strictly limited agricultural opportunities, so that rural dwellers, comprising nomadic Badu in the main, are small in number.

Group B: 30—40 per cent in Urban Areas

The nine states in this group (three of them in Northern Africa) have a variety of characteristics. Levels of urbanization are about the regional average. All (Jordan, Iraq, Syria, Cyprus, Lebanon, Iran, U.A.R., Tunisia, and Algeria) have important agricultural sectors and are at most 90,000 sq. km in area with the exception of Cyprus and Lebanon. Notably Iran and Algeria (1,648,000 and 2,382,000 sq. km in area respectively) are less urbanized than the group as a whole.

Group C: 20—30 per cent in Urban Areas

Containing South Yemen, Turkey, Morocco, and Algeria; these countries have relatively few obvious at-

Table 3. *Urban and city populations* a *for the Middle East in 1960*

Country	Percentage in urban areas	Percentage in cities
South-West Asia:	29.5	15.1
Israel	77.4	39.4
Bahrain	72.7	—
Kuwait	70.1	52.5
Qatar	60.0	—
Trucial States	40.0	—
Jordan	39.1	13.3
Iraq	39.2	21.4
Syria	36.9	26.4
Cyprus	35.9	17.7
Lebanon	33.1	26.8
Iran	32.9	17.3
South Yemen	28.0	23.5
Turkey	26.6	12.2
Saudi Arabia	6.2	13.5
Muscat and Oman	3.5	—
Yemen	3.4	—
Northern Africa:	29.6	18.2
U.A.R.	37.8	26.1
Tunisia	37.0	15.6
Algeria	31.2	16.4
Morocco	29.1	18.9
Libya	24.6	21.3
Sudan	7.3	2.7

a Statistics for the proportion of the population in cities are calculated using an arbitrary threshold of 100,000 people and over to define a city. Percentage in urban areas includes centres of all sizes which were *not* defined as "rural" by the country concerned. Calculated from Davis, K. [13].

tributes in common. They are all however, medium-sized (445,000—1,759,000 sq. km in area) and with dominantly agricultural economies. Libya's degree of urbanization is rising rapidly above levels shown here [15].

Group D: Under 10 per cent in Urban Areas

Sudan, Saudi Arabia, Muscat, Oman and Yemen are all strongly rural countries. They are also relatively "poor" and as yet are little touched by widespread industrialization. Towns in Saudi Arabia as in Libya, are growing rapidly in size and number as the effects of the oil industry are more widely felt.

Thus, it seems that Kuwait ranks not only amongst the most highly urbanized states in the Middle East, but also in the world (Tables 2 and 3). Inevitably, the comparison of Kuwait with states as large as Iran and Turkey needs some justification. Two points substantiate the parallels drawn between the very different groups of nations above.

First, in the world today, national boundaries are increasingly significant divisions of the earth's surface, which in our geographical studies we must recognize and use, despite the anomalies apparent in the status quo.

Secondly, when comparing e. g., Libya with Kuwait, the "effective state territory" of Libya (i. e. the Benghazi and Tripoli plains) is not much larger than Kuwait's effectively occupied territory. Subsequent analysis provides some more substantive reasons for drawing similar international parallels despite the problems involved.

4. Elements Peculiar to the Urbanization of the Middle East

Amongst the general factors responsible for the world-wide growth of towns, Hauser [16] cites the following factors:

I. A widespread increase in overall population since the Neolithic period.

II. Man's increased ability to regulate and control his natural environment.

III. A rise in the level of technical competence.

IV. Developments in social and political organization.

While these factors undoubtedly apply to the Middle East as much as to other parts of the world, they fail to explain the urban dominance of many cities of the region and the high level of urbanization in at least half of the Middle Eastern nations. Certainly Berry's [17] three propositions—that a state's size, its political immaturity and its economic simplicity are determinants of that state's degree of urban centralization—are particularly relevant to the Middle East, where all three attributes often coincide within a single state.

In addition to these factors, there are a variety of more specific reasons why Middle East cities are abnormally influential in the area. Fisher [18] mentions several, the main reason being that because of "a variety in geographical environment—rich oasis and coastal plain, mountain, desert, steppe and forest—there soon arose a diversity of economic production, and hence a need for exchange and market centres". He also cites the following:

I. The significance of the defence role, pointing out the number of tribal strongholds which later became cities—Aleppo, Ankara, Mosul, and Tabriz.

II. Various dynasties established "planted" towns in the Middle East, e. g. the Qajars and Teheran in 1788.

III. Small energetic groups seized power and ruled large territories from small urban bases, e. g. the Ottomans.

IV. All the religions of the Middle East have an urban origin; numerous large cities have grown up with this religious role as their major function—Jerusalem, Mecca, Medina, Karbala, Najaf.

V. Continued external contacts have been conducted with the outside world through the coastal cities since the Greek and Roman periods onwards, resulting in an unprecedented degree of urban economic dominance.

VI. Wealth is concentrated in the cities by the presence of absentee rural landlords within their bounds.

VII. Politically, Middle Eastern cities are dominant because government was, until recently, almost entirely concerned with the urban areas, to the detriment of the rural areas.

To these, we can add a further list of factors given by Shiber [19]. They include:

I. The recent devolution of political autonomy and the emergence of new nation states, each of which requires its national capital and its network of regional urban administrative centres. Post-war examples include Iraq, Egypt, Syria, Lebanon, Libya, Algeria, and Kuwait.

II. Oil discoveries have accelerated urbanization by focusing direct employment opportunities in oil camps and at tanker terminals, and indirect employment opportunities in the capital cities. Examples include: Iraq, Iran, Kuwait, Libya, Bahrain, Saudi Arabia.

III. As a result of the oil revenues, capital has been made available for investment—mainly in urban areas. Notable indirect beneficiaries from this flow of capital are Lebanon and Egypt.

IV. Inevitably the establishment and expansion of Israel has unsettled large proportions of the rural populations of Palestine. Many of these refugees—totalling approximately 750,000—took up residence in urban rather than rural areas,—such as Amman, Kuwait, Beirut, and Riyadh.

V. Following on from this, changes of regime have prompted sizeable numbers of people to migrate to urban areas. Frequent coups and the use of the military in government are spurs to movement, although not all of this movement is from the countryside to the cities.

VI. Since government has traditionally been city-based and city orientated, policies encouraging rural development have been sadly lacking. As a result, a great gulf separates levels of living in the towns and in the countryside of most Middle Eastern countries.

VII. Finally, the application of foreign aid, both civil and military inevitably focuses attention on the points of external contact with the aid-giving countries.

5. Conclusion

This Chapter has, in turn, examined the general theories linking urbanization and economic development on a world wide scale; narrowed this with a study of the application of these theories to the Middle East in particular; and concluded with a discussion of the elements which are unique within the Middle Eastern context. We have shown that the links between the two processes of economic development and urbanization are by no means as simple as some texts suggest. We raise the large question of whether urbanization and development are problems to be solved on a universal front or within a smaller regional context. Similarly, we question the validity of both cross-cultural and cross-temporal applications of general urbanization theories. At this stage no firm conclusions can be reached concerning these propositions. The foregoing study justifies the closer examination of Kuwait to answer these and related questions. It is apparent however, that while Kuwait itself has been regarded as an exceptional state internationally, it has important parallels within the Middle East area and beyond. In addition, its pattern of urbanization will have to be considered in the formulation of any general theory concerning the causes and progress of urbanization as a whole in non-Western contexts.

The task of the next and subsequent chapters will be first, to outline the physical and economic factors underlying the Kuwait urbanization, followed by a close examination of its progress, characteristics and implications.

II. The Urbanization of Kuwait

Urbanization is not a wholly continuous process; in the Middle East, certainly, as Adams [1] has shown for Iraq, phases of urbanization alternate with phases of "ruralization". This and the next Chapter trace both relatively constant physical elements and more fluid historical and economic changes which have affected urbanization in Kuwait. These themes are then followed through to the present day and identify historical and contemporary phases of urbanization.

In Arabia, periods of prosperity have alternated with periods of hardship and strife. It is imperative to isolate the origins of these phases to comprehend the present period of prosperity and its correlate, urbanization. As a general statement, it is possible to say that Kuwait's periods of prosperity have coincided with periods when external contacts were most prolific and widespread, while periods of depression occurred when Kuwait was restricted to the resource base within its own frontiers. What are the indigenous resources of Kuwait and what kind of setting does Kuwait provide for the growth of towns and cities?

1. The Environment

Kuwait is a small state with only recently defined boundaries (see Map 1). It covers an area of 16,000 sq. km of which 1,000 sq. km represents the offshore islands. To the south lies the Neutral Zone jointly administered by the Kuwait and Saudi Arabian Governments and with an area of 5,700 sq. km. Kuwait lies between 28° and 30° north of the Equator and between 46° and 48° east of the Greenwich meridian. Bounded to the north by Iraq, and to the west and south by the al-Hasa province of Saudi Arabia, Kuwait consists of a small triangle of land centred on the Bay of Kuwait.

This bay, extending 45 km westwards into the State, was until recently Kuwait's principal natural asset. Along the entire Gulf coast of Arabia, there is a dearth of good natural harbours between Basra in Iraq and Dubai in the Trucial States. The Bay of Kuwait, despite its shallowness and tendency to silt up (it is, in fact, a fossil river estuary), nevertheless provided a sufficient depth of water and harbour facilities for the small boats carrying on the bulk of Kuwait's coastal traffic. The Persian Gulf itself with an area of 240,000 sq. km has a mean depth of only 25 m.

Kuwait's surface lacks bold relief. In the west the land rises to just over 300 m above sea level, shelving gradually eastwards so that the eastern third of the State —including all the permanently inhabited districts—is less than 150 m above sea level.

The uplands to the west comprise gently undulating gravel plains whose monotony is broken by occasional knolls of sand collected around scrubby bushes called "hamdth" in Arabia (a generic word for saline brushwood), with "nussi" grass (*Aristida plumosa*) and "arfaj" scrub (*Rhanterium epapposum Oliv.*). Further details of the vegetation of the district are provided in Dickson [2] and Dickson [3]. Evidence of fluvial erosion is widespread but the most striking relief feature of the uplands is the great Wadi Batin trending SSW to NNE, along Kuwait's western frontier. On average 8—10 km wide with a relief as great as 70 m, the wadi forms the most marked feature of the western gravel plains.

Eastwards the gravel plains fade gently into an area of low relief, where the surface is mainly composed of windblown sand. Three notable relief features stand out from this almost flat landscape, the first of which is the Jal az-Zor escarpment running along the north shore of the Bay of Kuwait in an arc 80 km in length. Local relief reaches 130 m at Mutla' near the south-west corner of Kuwait bay where lower Fars formation limestones (lower to middle Miocene age) outcrop in a jagged ridge [4] (see Map 2).

Second, the Ahmadi ridge on which is situated the oil town of Ahmadi, stands out as a long whale-back rising to 115 m above sea level, paralleling the east coast and just 8 km inland. The feature is possibly the result of horizontal compression in post-Eocene times and is hence related to the Zagros orogeny. The Ahmadi ridge provides the necessary elevation for the gravity-feed storage tanks supplying the tanker terminals of Mina Al-Ahmadi, an important element in the low production cost of Kuwait oil. In addition, the ridge provides a slight cooling effect in summer as well as a pleasantly undulating site for the town of Ahmadi.

The hills at Wara, Burgan, Gurain, and Madiniyat in the southwest corner of the State provide the third physical feature of note in eastern Kuwait. Rising to about 30 m above the surrounding plain, these hills are composed of limestones of the Kuwait Group (Miocene-Pleiocene age), often capped with hard siliceous sandstone and chert which weathers to a dark brown colour.

Apart from the uplands of the south-west and the three groups of features mentioned above, Kuwait has a monotonously level surface which is covered only by a low scrubby vegetation. The climatic factors responsible for this scattered vegetational cover are important in that they also have a bearing on both human comfort and economic life (see Map 3).

Kuwait City (Fig. 1) in summer is among the hottest capitals in the world, while in winter, temperature minima approach freezing point. Equally extreme is the rainfall, which, as long as detailed records have been kept (since 1955), has never exceeded 200 mm annually. In several very dry years, less than 30 mm has been recorded. A little elaboration on these points is required before considering some aspects of the daily weather pattern.

Text Fig. 1 presents a summary of the available climatic data on Kuwait. Two seasons can be easily recognized with only short intervening periods of climatic transition (Table 4).

a) Summer

Between May and September, mean monthly temperatures exceed 30° C. More specifically, average night minima exceed 24° during this period, while monthly maxima are extremely high (over 37°). With temperatures of this order, relative humidity drops below a maximum of 50 per cent throughout the summer. Although the ambient air temperatures are extremely high —too high for strenuous out-door activity—the dryness of the air throughout the summer does facilitate body cooling. By comparison, less fortunate locations, such as Muscat town and Bahrain suffer both high temperatures and high humidities in summer.

Rain is never recorded but winds can be relatively strong during the summer. In August, 30.2 per cent of the winds exceed 11 knots compared with the January figure of 26.0 per cent. These are surface winds, very largely resulting from differential heating of the sur-

rounding land and water masses and have no vertical resultant, despite their great potential energy, because of a high level intrusion of air which lasts all summer, below which is created a stagnant zone of still air [5]. As the wind roses show (Text Fig. 1), the prevailing winds at all seasons are north-westerly, although in summer strong south-easterly winds are occasionally recorded, and bring with them the humidity experienced in the coastal areas further down (south-east) the Gulf (Text Fig. 1).

Clear skies, strong sunshine and the exclusion of the westerlies are responsible for the high temperatures in summer. While there is a difference of a few degrees between the coastal and the inland stations, the waters of the Gulf exert only a small moderating influence on summer temperatures. The explanation lies in the shallowness of the Gulf waters whose temperature rises to over 30° C in mid-summer.

b) Winter

Between winter and summer and vice versa, only very short periods of pleasantly warm and sunny weather intervene, which hardly warrant the titles of spring and autumn yet bring with them the colourful and welcome bird migrations following the Caspian-East African flyway. At the end of the summer, temperatures drop sharply from September onwards, while the rise in temperatures which takes place in May is equally sudden. Explanations of this phenomenon in terms of world air mass climatology can be found in Fisher [6], Trewartha [7], Banerji [8], and H.M.S.O. [9].

Winter is characterized by a greater unpredictability in both climate and daily weather than summer. Day to day variations in temperature, visibility and cloud cover occur in accordance with the movement of frontal disturbances through the Gulf. If we compare the January and July mean temperature maxima and minima at Shuwaikh over the 12-year period 1956—1967, it can clearly be seen that winter is a period of changeable weather compared with the monotony of the weather in summer. In January, the relative variability of the mean maxima was 39.3 per cent, while in July the relative variability was only 1.6 per cent. As for the mean minima, in January the relative variability was 62.3 per cent compared with only 1.0 per cent in July.

As expected with such very low total rainfall amounts, both seasonal and areal variability is large. Falls are local, and as Table 4 shows, can occur in any of the 6 winter months—even outside them, (e. g. in May, 1963). Coupled with the high rates of evaporation, such a rainfall provides an inadequate basis for even dry farming methods.

c) Daily weather

Summarizing, it is plain that only for relatively short transitional periods—"spring" and "autumn"—is the weather in Kuwait congenial to outdoor life. In midwinter, it can be extremely cold with occasional heavy downpours of rain that produce traffic chaos and widespread flooding. In addition, outbursts of cold northerly air lasting for several days called "shamals" can cause great discomfort indoors and out, not only because of the low air temperatures, but also because of the fine dust held in suspension in the air. This dust, penetrating every crevice in homes and offices, obscures visibility and

Table 4. Climatological data for Kuwait

All-readings are in centigrade		Jan.	Feb.	Mar.	Apr.	May	June	July	Aug.	Sept.	Oct.	Nov.	Dec.	Period over which records have been taken	Remarks
Ambient air temperature	Absolute maximum	26	33	39	42	48	48	49	49	46	42	37	31	1955—1964	Occurs between 1400 and 1600 hrs.
	Average maximum	18	21	26	31	37	43	44	44	41	35	26	20		
	Mean	13	16	20	25	31	35	37	36	33	27	20	15		
	Average minimum	8	10	14	19	24	26	29	28	24	19	14	9		Occurs between 0400 and 0600 hrs.
	Absolute minimum	−3	0	5	9	14	22	23	21	17	11	6	−1		
Dry-bulb temperature	Mean	13,5	16.7	20.1	24.9	31.1	36.0	37.2	37.1	33.0	27.5	20.1	14.9	1961—1964	In screen
Wet-bulb temperature	Mean	10.3	12.3	13.8	17.3	19.9	21.8	23.0	23.3	20.8	18.5	15.0	11.0		
Sun radiation	Absolute maximum	57	65	75	78	79	79	79	79	76	72	66	62	1958—1964	"Black-bulb" thermometer
	Average maximum	49	54	59	66	71	74	74	74	70	65	55	49		
Mean sea temperature		16	17	20	23	28	30	32	33	31	27	23	17	1955—1964	At the intake to distillation plant
Relative humidity (%)	Average maximum	75	74	69	63	54	40	43	44	48	59	71	80	1955—1964	Hygrograph
	Mean	63	53	47	43	36	27	28	29	31	38	51	61		
	Average minimum	43	33	26	24	18	13	13	13	14	18	32	43		
Average hours of sunshine/day		7.7	8.4	8.3	8.2	10.1	10.7	10.3	10.8	10.3	10.0	8.1	6.9	1955—1964	Campbell-Stokes recorder
Rainfall (mm)	1955	8.6	—	6.4	6.0	7.0	—	—	—	—	—	0.5	44.8	Yearly total 73.3	Tilting-Syphon self-recorder
	1956	10.8	4.7	6.0	14.2	—	—	—	—	—	—	—	119.3	155.0	
	1957	9.8	14.3	15.4	26.3	8.9	—	—	—	—	0.1	89.1	1.3	165.2	
	1958	12.1	0.7	7.6	2.2	1.9	—	—	—	—	—	15.7	61.7	101.9	
	1959	32.2	13.5	10.0	9.4	1.8	—	—	—	—	—	9.3	23.8	100.0	
	1960	7.7	2.6	2.7	4.2	—	—	—	—	—	—	11.3	0.1	28.6	
	1961	22.7	16.3	45.9	29.1	1.1	—	—	—	—	—	66.3	14.4	195.8	
	1962	27.1	3.2	4.5	18.9	0.1	—	—	—	—	—	0.1	12.7	66.6	
	1963	0.4	23.9	1.4	19.9	21.4	—	—	—	—	—	7.1	13.1	87.2	
	1964	12.2	2.2	2.2	—	—	—	—	—	—	—	1.1	8.6	26.3	

Except where stated, all readings were taken at Shuwaikh, Lat. 29°20′ N., Long. 47°57′ E.

causes severe disruption of communications. For example, at Christmas 1967, Kuwait airport was closed for a week due to one such dust storm.

By contrast, the summer season poses almost exactly opposite meteorological hazards. With high temperatures, clear skies, and hot breezes, evaporation both from the ground and from the skin is extremely high. Dehydration is a serious hazard amongst manual labourers working in the sun [10, 11]. Ground visibility is poor due to the shimmer of a persistent heat haze. All metal objects exposed to the sun become untouchable because of their heat. Car parks have to be roofed and cars themselves have to change to a thicker form of oil for the summer. Little refreshment can be gained from sea-bathing, because of the high temperatures and salinity of the Gulf. While the summer is undoubtedly an uncomfortable season in Kuwait and one in which Kuwaitis and non-Kuwaitis alike try to avoid by migrating temporarily to Lebanon or Europe, the discomforts of the summer have been greatly relieved by the widespread use of air-conditioners. Kuwait claims to be the most extensively air conditioned world capital for, in summer, electricity consumption is at its peak because of the air-conditioning units. Houses, offices, shops, hotels, restaurants, and many cars are fully air-conditioned. Walking around the streets in summer, the buzz and whine of the air-conditioners are a feature of almost every building one passes.

Having described the physique and climate of Kuwait, we will now be concerned with the compounding of these elements to form the physical resources available

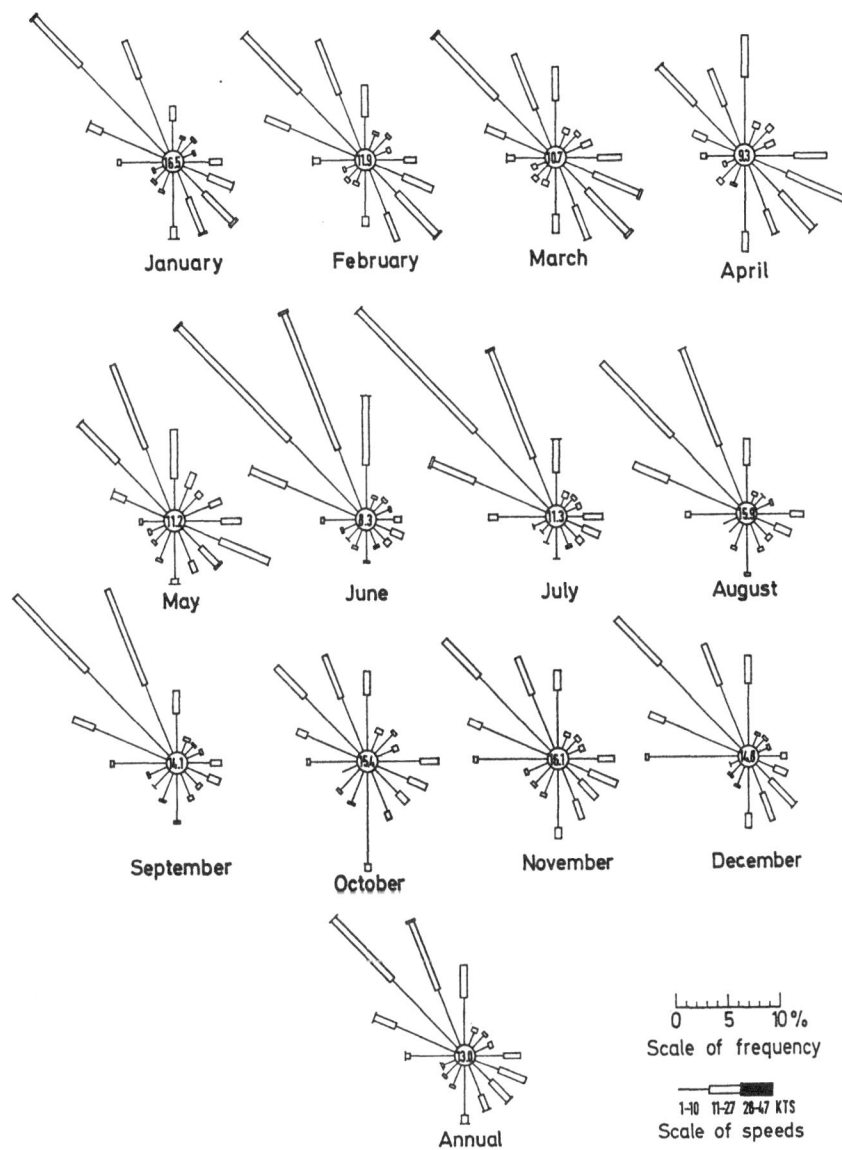

Text Fig. 1. Monthly percentage frequency of winds at Kuwait international airport

to man, by a survey of Kuwait's water resources, fundamental to all economic life, moving on to a consideration of the potential for agriculture, fishing, and industry.

2. Water Resources

a) Occurrence

Four sources of potable water are known to exist in Kuwait. They are:

I. Water occurring in scattered shallow well-groups in the eastern half of Kuwait. This water, derived from percolation in hollows during rainy periods, has been used for long periods by the Badu at such places as Jahra, Subaihiya, and Tawil. It becomes saline and almost undrinkable in late summer.

II. Water occurring in the Dibdibba formations about 30 m below the surface. This fossil water is fresh, and was first discovered in 1960 at Raudhatain in the north of Kuwait near the Iraqi border. The Raudhatain reservoir was estimated as containing at least 100 billion gallons in 1961.

III. A third deeper source of ground water is that occurring in the Dammam limestone which outcrops at Ahmadi and the dips northeastwards at 1.7 m/km. The static fluid level falls in the same direction at less that 0.8 m/km but unfortunately the proportion of total dissolved salts rises sharply from its lowest value of 500 parts per million in the south-west to over 1,000 p.p.m. nearer Kuwait City [12].

IV. A final source of water is the sea. Distillation began at Mina Al-Ahmadi in 1950 and since then, rapidly increasing volumes have been produced by the distillation plants of both oil companies and the government. In 1966, over 58 per cent of the water used in Kuwait came from the numerous distillation plants [13].

b) Development

Up to about 1925, the population relied entirely on the brackish water occurring in shallow wells in eastern Kuwait. Within Kuwait City, a few groups of wells produced potable water but most of the city's population relied on water from Shamiya and Hawalli outside the Old City wall. With an increased population after 1920 (see below) these shallow wells with their seasonal increase in salinity provided insufficient water for the city. The nearest available source of potable water was the Shatt al-Arab (lower Tigris-Euphrates), so that after 1925, a sea-borne trade in fresh water began which lasted until the opening of the first distillation plant in 1950. Initially, water from the Shatt was scooped up in goatskins and barrels. By 1939 the trade had become important enough for a water carrying company to be formed with a fleet of several dhows. 35 dhows were engaged in this trade by 1946, and three further ships were privately operated. Even the Kuwait Oil Company (K.O.C.) was supplied with this water which was distributed by donkeys and motorized scooters. Its selling price was 20 fils per gallon (20 fils is approximately equal to 2½ p. sterling). In the peak year of 1947, 80,000 gallons of Shatt water were reaching Kuwait daily [14].

By 1950, demand by K.O.C. and the growing population of Kuwait City had far outstripped supplies from the Shatt and the shallow local well sources. Sea water distillation was begun at Mina Al-Ahmadi (Map 1) to meet these expanding demands with a daily production of 600,000 gallons. First 80,000 gallons and then 250,000 gallons were pumped the 40 km across the desert to Kuwait, until in 1953, the Government opened the first of several new destillation plants at Shuwaikh, just west of Kuwait City. Capacity rose from 1 million gallons daily in 1953 to 6 million gallons daily in 1960. Present capacity is 8 million gallons daily which will rise in the near future to 12 million gallons (Text Fig. 2).

In addition to the Shuwaikh desalinization facilities, sea water is also distilled at Shuaiba, the Government's new port and industrial estate in south Kuwait (Map 1).

There, 3 million gallons of fresh water are produced daily while a further capacity of 2 million gallons is under consideration. Present desalting capacity totals 11 million gallons per day but a projected capacity for 1972 in the Five Year Plan is 42 million gallons daily [15].

Parallel with this rapid expansion of distillation facilities has been an equally rapid development of the brackish water resources at Sulaibiya, 27 km south-west of Kuwait. Some of this water, called simply "Sulaibiya" in Kuwait, is blended with the distillate to improve its overall taste, and the rest is distributed in tankers for use in sewers, in gardens, and for agriculture. At present, most houses in Kuwait have both brackish and fresh water supplies delivered to them by tanker lorries. Some drinking water is still delivered by motorized scooters.

Text Fig. 2 shows the rapid rise in the production of Sulaibiya water, in 1967 totalling 12 million gallons daily [16]. Particularly noteworthy is the rising demand for both brackish and fresh water in summer.

Overall, there is a heavy reliance on desalinization to compensate for the deficiency of rainfall and ground water. Although the search for fresh water continues particularly in west Kuwait, Raudhatain, the largest fresh water reservoir so far encountered, supplies only 4 million gallons per day—a rate which can be sustained for 20 years. A significant development is the plan to lay a water pipeline from the Shatt al-Arab to Kuwait. The contract with Iraq was signed in 1964 and an international panel was appointed in 1965 to supervise the implementation of the agreement to supply Kuwait with 120 million gallons daily. 70 million gallons of this were ear-marked for agriculture [17]. For political and other reasons, work has not begun on the pipeline.

3. Power

Kuwait has an immense surplus of energy: oil exports in 1969 alone totalled 129.5 million tons, equivalent to approximately 53,542 million therms. Within Kuwait, the major energy source is natural gas which is produced as a by-product of oil. In 1966 for example 456,761 million cubic feet of gas was produced of which only 122,658 million cubic feet could be used [18]. Gas under pressure is used to generate electricity and to distill sea water in large dual-purpose plants at Shuwaikh and Shuaiba. In 1966, annual production of electricity exceeded 1,000 million kilowatts compared with a production of only 87 million ten years earlier. This electricity is sold for 2 fils per kilowatt—an almost negligible amount—resulting in a very high per caput consumption, especially in mid-summer when air-conditioners are in widespread use.

4. Agriculture

Lack of rain and underground water coupled with high evapotranspiration rates severely circumscribes agricultural activities in Kuwait. Recent attempts to extend cultivation beyond a "garden-culture" are a noteworthy addition to the traditional pastoral economy.

a) Farmers

Field cultivation was until recently almost unknown in Kuwait. Early this century, Jahra was the only village with enough water to produce even dates and *jit* (a form of alfalfa). Lorimer [19] recorded only 2,000 date palms. Some of the villages in the east coast had a few date palms but from the very earliest period, Kuwait has been a net food importer.

In recent years, steps have been taken to ease this heavy reliance on imported food—particularly vegetables and milk. A Government-sponsored experimental farm was established in 1953 and now covers 40 hectares on the southern perimeter of Kuwait City. Initially it produced trees and shrubs for Kuwait's parks and traffic islands but now it has sections dealing with poultry, a dairy, and a nursery. Salt resistant strains are being tested and evolved on the farm [20].

A new departure is the establishment of a large-scale hydroponics section in association with several Japanese experts brought to Kuwait as part of a contract with the Japanese-owned Arabian Oil Company. A soil survey of Kuwait is in progress but as in other Gulf states, the limiting factor in agriculture remains water and not soil [21]. The relative insignificance of agriculture in Kuwait is brought out by the statistics of the 1965 Establishment Census; only 559 workers contributed just 0.4 per cent to the total Gross Domestic Product formation in 1965—6 [22].

b) Nomads

In 1957, 15,679 desert dwellers or Badu were recognized in the Census compared with 6,187 in 1965.

Lorimer at the beginning of the century put the number of desert dwellers within Kuwait territory at 13,000 [23]. While these Badu are subject to large enumeration errors because of their constant movements across international boundaries, the figures reflect a generally accepted trend. The Badu population is being gradually reduced, and with it, the production of livestock and livestock products.

Exports of sheep, goats, and camels were important up to World War II, but since then have dwindled to very low levels. In 1966, exports of livestock were worth less than K.D. 2,000 and the Planning Board estimated that less than one-tenth of Kuwait's meat supplies were met from the desert herds [24]. The measure of Kuwait's meat imports was emphasised in 1965 when a Yugoslav ship carrying 5,000 live sheep from Australia capsized while tying up in Shuwaikh harbour.

5. Fishing

a) Pearling

As Blegvad and Loppenthin [25] showed the fishing resources of the Gulf are considerable, but it is for pearls rather than fish that the waters of the Gulf have been traditionally farmed. Lorimer's account of the pearl industry before oil remains the most authoritative although many authors have described the industry since [26—30].

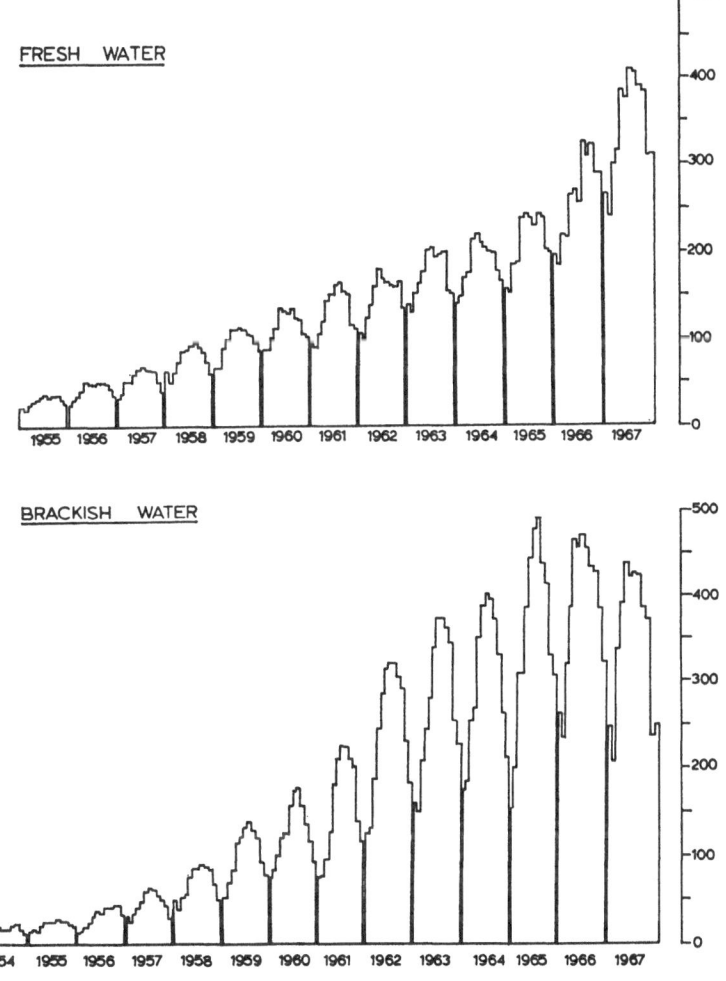

Text Fig. 2. Water production in Kuwait in millions of imperial gallons 1954—1967.
Note the steep rise in water consumption in summer

In 1905—6, Lorimer estimated the value of the pearl catch for the whole Gulf at £ 1,434,399 at the very lowest. Kuwait's income from pearling he put at 134,700 Indian Rupees. The industry as a whole occupied 9,200 Kuwaitis in 461 boats. In the inter-war period, pearling suffered a serious setback not only because of the severe depression of the European markets but also because of the advent of the Japanese cultured pearl. Some recovery occurred in the post-war period but today the industry has dwindled into insignificance, though still retaining some interest for lovers of pearl jewelry, students of maritime history, and those concerned with the lasting physiological effects still to be seen in ex-pearl divers.

b) Fishing

Besides the catches of fish landed locally by small fishing boats, three large commercial companies are responsible for most of the fish landed. All three companies fish for prawns which are then frozen and flown to North America and Europe. Operating since 1959, the Gulf Fishing Company has 63 vessels of which 59 are trawlers. The Kuwait National Fishing Company has 42 boats and two mother ships and 10 trawlers.

While fishing is important as a means of industrial diversification, the industry in 1965 occupied only 1,258 men [31].

6. Mineral Resources

Industries

Kuwait lacks manufacturing industry either on the bazaar level or as an organized factory system. Lorimer noted an absence of craft industry 60 years ago and only recently, with the sizeable resources of cheap power and chemical feedstock, has some attempt been made to combine these elements in a chemical industry. A petrochemical complex is being created at Shuaiba in south Kuwait including an oil refinery with a throughput of 95,000 barrels per day in 1968. The Kuwait Chemical Fertilizer Company will eventually be producing daily 400 tons of ammonia, 550 tons of urea, 400 tons of sulphuric acid, and 510 tons of ammonium sulphate. As a by-product of sea water distillation, the Shuwaikh plants produced 3,700 tons of sodium chloride, 930 tons of chlorine, 1,240 tons of caustic soda, and 126,000 gallons of hydrochloric acid in 1966 [32].

Two other industries are associated with the building trade. The first, the manufacture of sand-lime bricks in 1966 produced 49 million bricks. The second, manufacture of unslaked lime, produced 1,320 tons in 1966.

Kuwait has no other manufacturing industries of note, with most of the employment opportunities in the service and food processing industries. In 1965 over half

Table 5. *Distribution of the labour force by main economic activities, 1965*

	Kuwaitis	Others
Agriculture and fishing	1.4	1.0
Mining and quarrying	3.4	4.1
Transformative industries	4.5	11.6
Building and construction	3.1	19.8
Electricity and water	4.1	3.8
Trade	12.8	12.9
Transport and communications	6.6	5.3
Government and services	63.5	41.0
Others	0.6	0.5
Total	100	100

Calculated from Census of Establishments, Table 2 (1965) (Arabic).

Table 6. *Per cent contribution of the various sectors to the gross domestic product*

Sector	1965—6	1966—7
Agriculture and fishing	0.3	0.4
Mining and quarrying	63.4	61.4
Industry	3.2	3.5
Construction	4.3	4.8
Electricity, water, and gas	2.1	2.3
Transport, storage and communication	2.8	2.8
Wholesale and retail trade	8.0	8.1
Finance, insurance and real estate	0.8	0.9
Housing	4.7	4.8
Public administration and defence	5.5	5.8
Services	4.9	5.2
Total	100	100

Source: Planning Board, 1968. Table 8 in: Economic Survey, 1966—7.

of the labour force were employed in government and services (Table 5).

Further, as Table 6 illustrates, manufacturing contributed only a tiny proportion (under 4 percent) to the Gross Domestic Product in both 1965—6 and 1966—7.

Virtually all other sectors are dwarfed by the mining and quarrying division, with industry contributing only 3 per cent to the Gross Domestic Product.

Our conclusion is that Kuwait's resource base is both difficult and narrow. Few states have such severe climatic regimes and such serious problems of water shortage. Economic opportunities have thus been more restricted than in other states of the Levant and North Africa, so that for these reasons, Kuwait's economic history has followed a course different from that of the Middle East's as a whole. The Bay of Kuwait and the State's strategic location at the head of the Gulf are the principal physical assets on which Kuwait has been forced to capitalize.

III. The Early History of Kuwait

As a waterway linking Mesopotamia with the Indus Valley, the Persian Gulf was the scene of some of the earliest known voyages in the pre-Christian era. It was here that one of the world's great trade routes developed, and perhaps it is not surprising when one reviews the inhospitality of its shores that it was the islands of the Gulf which predominated in the early phases of seaborne commerce, among the better known being Bahrain,

Kharg and Hormuz. The evidence that Kuwait also had its portion of this considerable trade comes from the interesting excavations on Failaka Island in the mouth of Kuwait Bay. These have revealed Bronze-age dwellings suggesting the existence of a civilisation, probably Sumerian, some three thousand years ago. These people, together with the Elamites, the Egyptians, the Hittites of Anatolia and the Harrapan peoples of the Indus River Valley were the most ancient of all known civilisations. They were a non-Semitic people who originally occupied South Babylonia but gradually spread down to the Gulf to form a strong link with the Indus Valley nations. The evidence for this is clear from the excavations at Ur (of the Chaldees) just over Kuwait's northern border and at Mohenjo-Daro in modern Pakistan. The remarkable burial mounds on Bahrain Island are also thought to have belonged to these early mariners whose artefacts can still be seen today stretching over nearly 2,000 miles, from Iraq to India.

Failaka Island later became the flourishing Greek colony of Ikaros in the fourth century B.C. It was then a wooded island, watered from many sweet-water wells and there is still considerable evidence of the quality of the colonists' life there which is recorded on tablets describing the visit of Nearchus, the Admiral of Alexander the Great, who sailed his fleet from the Indus River to the mouth of the Shatt-al-Arab in 325 B.C.

The Greeks apparently colonised the island as a defence against marauding Arabs, who alone among people in that area had failed to pay Alexander the respect he considered due. Eighteen centuries later the Portuguese also found it necessary to build similar defence posts, the ruins of one of these is still to be seen in Quaim Island in Kuwait Bay.

In these pre-Islamic times there were few, if any, large settlements on the coast of the Gulf other than those of the roving Arab pirates and the three trade and fishing ports of the site of modern Bushire (Abu-Shahr), Bahrain and El-Qatif, probably the site of ancient Gerra. At the tip of the northern arm of Kuwait Bay lay the ancient settlement of Kathima. Recent excavations with the finding of stone tools suggest that this spot was the home of a very early people. It remained for many centuries a thriving trading and military post, and in the early Islamic period it was often mentioned in Arab accounts as one of the main strategic posts of the Muslim armies. It was here that one of the closing battles of the Islamic conquest, the "Battle of the Chains", was fought between the Arab and the Persian armies in A.D. 636.

In contrast to the relatively sophisticated manner of living in the Greek colony of Ikaros, Arrian [1], historian of Alexander's expedition to India, had this to say about the simple and severe life of the people who lived on the barren shores of the Gulf. "Some of them indeed, make nets, of two stadia in length, and make use of the inner rind of palm trees, which they twist together, as we do hemp. But when the tide falls away, and the sea leaves their shores, they find vast plenty of fish in the small gullets, or hollow places, were the shore is not quite flat ... the small ones ... they immediately eat raw, the larger and tougher sort they lay in the sun to roast, and afterwards rub them to powder and make bread thereof, and some mix this powder and the flower of the wheat together ... but those who inhabit the most desolate parts, which produce neither trees nor fruits, live wholly on fish. They built their houses in this manner: the richer

sort among them, gather up the bones of whales or such other large fish as they find them cast upon the shore, and use the smaller bones for rafters, and some of the larger size for door-posts; but the people of inferior rank, build with the ribs, and backbones of other fish." Compare this with what Carsten Niebuhr [2] wrote of his visit almost exactly two thousand years later: "There is a striking analogy between the manners ascribed to the ancient Ichthyophagi and those of these Arabians. They live nearly all in the same manner, leading a sea-faring life and employing themselves in fishing and in gathering pearls. They use little other food but fishes and dates and they feed their cattle upon fish."

During the past two centuries many European travellers have mentioned Kuwait either from personal experience or hearsay. Carsten Niebuhr was one of the earliest of these. A German, he was the only survivor of the ill-fated Danish expedition to the Yemen which set out in 1761. He passed through Kuwait in 1765 and made no bones about his dislike of the climate when he described the terrible north wind which brings the sand with it: "The effects of the famous Sam Shun, or Samiel are instant suffocation to every living creature that happens to be within the sphere of its activity, and immediate putrefaction of the carcases of the dead." Niebuhr must have felt very low at that time, for the comparative merits of Kuwait's climate were well known to the people elsewhere in the Gulf, particularly Basra and Bushire. It had the reputation of having the mildest summer of any town on the coast, for the relatively cool west wind blew throughout the night and its health-giving climate and lack of mosquitoes was one of its early assets [3].

With the establishment of the Abbasid Caliphate in Baghdad in A.D. 750, trade with the Orient, India, the Moluccas (Spice Islands) and China, began to develop the pattern which led to the struggle for maritime supremacy among many nations in the ensuing millennium. Much of the trade was coastal and followed the Persian shore; in successive periods, emporia such as Siraf, Qais and Hormuz achieved widespread renown for their size and prosperity. The mariners of the Arabian shore probably shared to some extent in this trading prosperity, for several of the Arab geographers of the period mention Bahrain, the ancient Dilmun of the Assyrians, the great Semitic nation whose zenith was reached in the twelfth century B.C. Nasir-i-Khusran [4] visited Bahrain two thousand years later, in A.D. 1051, and described its pearl industry and flourishing date gardens, also eulogised by Ibn Batuta [5] as "a fine and considerable city with gardens, trees and streams". References by other writers to tribal instability and feuding along the Arabian coast at this period suggest that commerce was only a thin veneer overlying the traditional Badu economy [6].

The Portuguese phase of commercial monopoly in the sixteenth century largely passed by Kuwait, although they did establish two defensive positions, one on the island of Quam in Kuwait Bay, the other at the eastern tip of Failaka Island, but only traces of either remain. Of the succeeding activities of the British East India Company more will be heard later, but here began the tenuous thread which has linked Kuwait to Britain in friendly relationship for two hundred years.

As the importance of the settlement of Kathima diminished in the century following the Battle of the Chains in A.D. 636 there is little reference again to

Kuwait's fortunes until they took a turn for the better in the seventeenth century, when Al Kut, (the Fortress) was established as a summer residence of Barrak, a Shaikh of the Bani Khalid tribe, who observed grazing rights over the area we now know as Kuwait. The Bani Khalid were also in control of all trade into Central Arabia as well as most of the harbours, such as they were, in Eastern Arabia. These Bani Khalid belonged to the Rabi'a, an Adnani tribe from the Nejd and it was the 'Utub portion of the tribe which migrated from Al-Aflaj in Central Arabia because of drought conditions to settle in Bani Khalid land where the town of Kuwait stands now. We know that the founder of Al-Kut, Shaikh Barrak, died in 1682, and it was some time later in 1716 when the 'Utub migration began. The point of real development however seems to have coincided with the death of the local Bani Khalid Shaikh, Sulaiman al Hamid in 1753 and the election of his successor from the 'Utub, Sabah bin Jabir, the first of the long line of the present ruling family of Kuwait. This development arose from the clever consolidation by the 'Utub of their relationship with the Bani Khalid through marriage. Building on the Bani Khalid's established trading connections, the 'Utub were able to expand both seaborne and caravan trade. The first fifty years of Kuwait's foundation was marked by a "high degree of prosperity" [7].

This contrasted with the low ebb of commercial prosperity in the Gulf as a whole during the second half of the eighteenth century, mainly due to the Turkish-Persian rivalry, the Persian Civil Wars and an upsurge of piracy. Kuwait fared well because of the Persian occupation of Basra between 1775 and 1779: this had a lasting effect: "... the prosperity at this time was considered to stand in an inverse ratio to that of Basra. A noteworthy consequence of the Persian occupation of Basra was the migration of a number of merchants to Kuwait and the removal of others who did not feel themselves secure even there, from Kuwait to Zubara in Qatar. The trade and general growth of both Arab seaports was strongly stimulated by these events ..." [8].

Kuwait's importance as a caravan transhipping centre between Aleppo in Syria and Bombay derived from the harbour facilities in the Bay. Caravans often numbered five thousand camels and a thousand men. The journey would usually take about seventy days between Aleppo and Kuwait although mail riders could do the distance in fourteen. Thus it was that these three factors, the continual nervousness in Basra, the attraction of a deep water harbour and good transhipping facilities and rapid mail transit to Aleppo combined to recommend Kuwait to the British East India Company. At this time the town of Kuwait contained over ten thousand people, no mean size.

In May 1776 a skeleton staff of the East India Company together with equipment and merchandise transferred from Bushire to Kuwait after having been evacuated from Basra the previous year. Their arrival in Kuwait had been facilitated by the visit of H.M.S. "Dolphin" and H.M.S. "Seahorse" which surveyed the Bay of Kuwait for the first time and stood by to see the East India Company's men installed. It is of interest that one who was later to become a very great Englishman, Horatio Nelson, was then a mid-shipman of eighteen in H.M.S. "Seahorse" and was thus one of the first of his countrymen to set foot in Kuwait [9].

The previous year the first postal dispatch to London had been made through Kuwait in a letter dated July 15th 1775 from Moore, Latouche and Abraham, the East India Company agents (at Bushire), to the Court of Directors in London [10]. The Company's factory was not permanently established because the occupation of Basra by the Persians ceased in 1779, but the mails continued to be landed at Kuwait to be carried to Aleppo. The salubrious air of Kuwait was frequently sampled by Company staff from Basra, none more important than Harford Jones, later Sir Harford Jones Bridges, H. M. Ambassador to the Shah of Persia, who was the factor at Basra and summered at Kuwait in 1790. In 1793, when it was the Turks who this time were embarrassing the Factory in Basra, Harford Jones transferred it to Kuwait and set up a more permanent establishment which again remained for only two years perhaps because of the complaints of its staff that the supply of water was "infamously bad in quality being at once salt, sweet and bitter" [11].

One of the most controversial writers on Kuwait, Ralph Newins [12] has given an excellent appraisal of the significance of these early contacts between John Company and the Shaikh of Kuwait: "The Shaikh's act of hospitality in 1776, in allowing the Royal Navy to survey the Bay or Kuwait and a site for a factory, laid the small but vital foundation stone of a new nation as surely as George Washington simultaneously launched a new nation on the other side of the world. A sequence of events was set in motion which reached its logical conclusion in 1961, when Kuwait at last threw off the final traces of foreign interference and took her place in the world as a fully independent state ... an atlas and the pages of history show that Kuwait is the only place in the Middle East which has not changed hands since 1776. It has never been occupied by a foreign power, let alone been conquered, nor has it become a protectorate of the powerful nations with whom its destiny was involved, the Saudis, the Turks, the Egyptians or the British."

Carsten Niebuhr was the first of several European travellers who visited and described Kuwait. Another was Baron Stocquelor [13] who was there in 1831: he described it as a city of at least ten thousand with an armed population of five to seven thousand men. Trade amounted to 600,000 Maria Theresa dollars, 500,000 dollars in imports, on which the Shaikh levied a duty of two per cent. Apart from this levy there was little other government interference in the activity of the merchant community. At this time the Kuwaitis are said to have owned 15 large ships called "baghlahs", 20 smaller vessels ("battils") and about 150 smaller craft. Some of these were involved in the pearl industry of which the total value was 1,000,000 Bombay rupees for the whole Gulf. Bahrain was the main pearling port however, and Kuwait prospered more because it was a centre for smuggling goods into Turkish Iraq as well as a means of conveying the plunder of the Qasimi pirates from Bahrain to Persia. Troubles in Hasa also contributed to the flow of goods through Kuwait. This was by no means the last occurrence of smuggling activities in Kuwait which have continued regularly through the succeeding 150 years, being high-lighted by the blockade of Kuwait by the Royal Navy during the 1914—1918 war to prevent arms supplies to Turkish Iraq. Even today the Government of Kuwait maintains a camel corps to frustate this trade across its borders, particularly since the introduction of alcohol prohibition in 1964.

Kuwait continued to grow in status throughout the 19th century although direct sources of information are still scattered and thin. W. G. Palgrave, brother of the better known poet and anthologist, travelled extensively in Arabia. Although he never visited Kuwait personally, he wrote of what he had heard of the city at mid-century and had these pleasant things to say of the people of Kuwait:

"Among all the seamen who ply the Persian Gulf, the mariners of Koweyt hold the first rank in daring, in skill, and in solid trustworthyness of character. Fifty years since, their harbour with its little town was a mere nothing: now it is the most active and the most important port of the Northerly Gulf, Aboo-Shahr hardly or even not excepted. Its chief, Eysa, enjoys a high reputation both at home and abroad, thanks to good administration and prudent policy; the import duties are low, the climate is healthy, the inhabitants friendly, and these circumstances, joined to a tolerable roadstead and a better anchorage than most in the neighbourhood, draw to Koweyt hundreds of small craft which else would enter the ports of Aboo-Shahr or Basra ... In its mercantile and political aspect this town forms a sea outlet, the only one for Jabal Shammar, and in this respect is like Trieste for Austria" [14].

Kuwait remained a tiny, insignificant Shaikhdom throughout most of the 19th century, but towards the end political events in the surrounding territories drew it first within the Ottoman sphere of influence after the Najd expedition of 1871 [14] and later within the British.

One of the first intrusions of the outside world into Kuwait had begun in the 1870's when the British India Steam Navigation Company steamers began calling there. The service was suspended however, when Kuwait's prosperity was thought to be prejudicial to that of Basra and it was not until 1901 that a regular weekly service was introduced. Kuwait's status within the Ottoman Empire was never entirely clear since the Shaikhs retained a degree of autonomy while accepting the Ottoman title of "qaimaqam" over the territory as a confirmation of their succession. British policy in the lower Gulf, while aimed at suppressing piracy and the slave trade at sea, had at the same time been to avoid interference with internal affairs on the Arabian mainland throughout the 19th century.

But two acts drew British attention to Kuwait towards the end of the period; the first was Shaikh Mubarrak's fratricide and seizure of control in Kuwait followed by growing acts of piracy in the Shatt al-Arab by Arabs apparently under Mubarrak's suzerainty. At first wishing to hold the Ottoman Porte (Government) responsible, the British performed a remarkable political volte face when Count Kapnist, a Russian subject, applied to the Porte in 1898 for premission to build a railway from Tripoli in Lebanon through Baghdad to terminate in the Persian Gulf at Kuwait. Shaikh Mubarrak was promptly asked to sign an Exclusive Agreement promising not to "receive the Agent or Representative of any power or Government at Kuwait ... without the previous sanction of the British Government; and he further binds himself, his heirs and successors not to cede, sell, lease, mortgage, or give for occupation or any other purpose any portion of his territory to the Government or subjects of any other power ..." [15].

Originally conceived to oppose Russian expansion in the Gulf, the Agreement proved equally effective in thwarting later German plans to construct a railway terminal in Kuwait, for which purpose a German survey team reached there in 1900.

Growing British involvement in the Gulf prompted the despatch of J. G. Lorimer of the Indian Civil Service to the Shaikhdoms on the Arabian shore; he later produced his monumental two-volume "Gazeteer" in 1908 and 1915 [16]. For the first time, detailed information was available on Kuwait's history, population and economic life, and for this reason it is possible to trace Kuwait's development in the twentieth century with remarkable accuracy (Fig. 3).

With expanding trading connections and a forceful leader in Shaikh Mubarrak, Kuwait was prospering in the early years of this century. In Dickson's eyes, Kuwait became "to the Arab mind, a most attractive place to live in and the population of the town had nearly doubled itself" [17]. Most of the credit for this prosperity goes to Mubarrak, ruler from 1896 to 1915, "who really raised Kuwait from a place of little importance to a flourishing principality". Mubarrak was responsible for the provision of customs and warehousing facilities which were instrumental in increasing trade, but his demands for higher import duties and a tax from householders on his lands caused dissatisfaction locally.

Lorimer provides the first accurate information on Kuwait's economic life. He wrote:

"Pearl fishing is the premier industry of the Persian Gulf; it is besides being the occupation most peculiar to that region, the principal or only source of wealth among the residents of the Arabian side. Where the supply of pearls to fail, the trade of Kuwait would be severely crippled while that of Bahrain—it is estimated—would be reduced to about one-fifth of its present dimensions ..."

Bahrain was responsible for most of this catch, however, as Table 7 shows:

Table 7. *Value of the pearl catch by principalities in 1905—6*

Principality	value of catch in rupees
Bahrain	12,603,000
Trucial Oman	8,000,000
Lingeh	695,861
Kuwait	134,700
Masqat	22,500

Source: Lorimer, 1915, p. 2253 [16].

But in Kuwait, 9,200 men were involved in pearl diving, prompting Lorimer to remark: "The lower and middle classes of Kuwait almost all live by seafaring occupations, such as fishing, pearl diving and the coasting trade ..."

Local industries were confined to shipbuilding and the service and handicraft trades. All the materials for ship building were imported—the ribs from Karachi, the nails from India, and the rope and fibre from Calicut. Altogether 300 carpenters were employed in the production of 20 to 30 vessels annually. It is apparent from this that Kuwait was involved in a sizeable entrepot trade with Najd, Mesopotamia, and India and was prospering in its role as the only port serving Najd which was not under

Turkish control (Fig. 4). For this reason, arms were the most valuable import in 1905—6. Of Mubarrak's estimated income of 399,000 Maria Theresa dollars, the largest single item was the revenue from the sea customs amounting to 150,000 dollars.

Kuwait continued to develop on the basis of its overseas trade and income from pearling up to World War I, and of these the long distance carrying trade, was growing in significance. While it prospered on these traditional industries, the lack of indigenous natural resources laid the state open to the foibles of international politics. Despite British attempts to control the lucrative arms traffic in the Gulf (not, lest it be thought, for itself), Kuwait amongst other Gulf ports was deeply involved in the supply of arms and supplies to both sides during the First World War. Thus in 1918, the Royal Navy enforced a sea blockade on Kuwait, the first of several embargoes which severely curtailed trade and prosperity. Furthermore, during the subsequent 'Ikhwam rebellion in Arabia, 'Ibn Sa'ud was concerned about the supplies reaching his opponents through Kuwait so that he too enforced a strict blockade on trade between Najd and Kuwait from 1923 to 1937. Dickson [18] describes this period for Kuwait as "a long-drawn-out fourteen years' agony" and Zahra Freeth, (his daughter) characterizes the period as one of economic ruin [19]. Fraser's report [20] on the other hand provides an excellent contrast by describing the pre-blockade period thus:

"Kuwait is growing rapidly and has spread far outside its old walls. Its trade is steadily increasing and the Shaikh is waxing rich. He is reputed to be able to put 10,000—15,000 fighting men in the field."

Despite the British attempts to have the Saudi embargo lifted, Kuwait's economy remained in the doldrums until 1937. Threatened by the incursions of the 'Ikhwan, culminating in the Battle of Jahra in 1920 and the rapid construction of the city wall in 1920 (Figs. 5 and 6) Kuwait was faced with a major economic crisis due to the intrusion of the Japanese cultured pearl on to

the world markets, coupled with the financial crises in Europe and North America in the late 1920's and early 1930's. But in December 1934, Shaikh Ahmad signed a salutary document; the Agreement granting a concession to explore for oil to an Anglo-American concern, the Kuwait Oil Company. Despite the payments of dead rents and the discovery of oil at Burgan in 1938, the Second World War ushered in a further period of economic hardship for Kuwait. Oil operations were suspended between 1942 and 1945 because of lack of supplies and trade was interrupted by the occupation of many of the Gulf ports by the British as a measure to protect their supply lines to Russia and the Turks.

Dickson remained in Kuwait during the War to oversee the Oil Company's installations, having retired as Political Agent. His remarks are a telling reminder of Kuwait's prosperity from the earliest period recorded, has depended heavily on trade and external contacts:

"It (wartime) was a difficult time for everyone, especially the poor of the town and in the desert, for food, clothing, and medicines were almost unobtainable, and great distress prevailed ... my wife and I managed privately to import by sailing boat from Persia a hundred pairs of grinding-stones, which we issued to the starving Badu of the hinterland (one set to every ten tents), to enable them to grind barley which we also bought for them ... At that time there was practically no wheat or flour in the town, rice was terribly scarce, and the price of dates had reached a starvation level" [21].

With the end of hostilities in 1945, further drilling and oil production were quickly resumed, so that by June, 1946, Kuwait exported its first shipload of oil (Fig. 7). Phases of prosperity in Kuwait have been associated with the period when international trade in the Gulf area has been at its height. Throughout the period up till 1946 Kuwaitis relied heavily on external contacts and had been prepared to make the most of their one major natural resource until then, their geographical location.

IV. The Economic Development of Kuwait

Kuwait's economic development is not a history of steadily increasing wealth and prosperity. The State's dependence on the entrepot trade of the Gulf and eastern Arabia laid it doubly vulnerable to political and economic factors largely beyond Kuwait's control in both maritime and territorial realms. Just as the prosperity of the merchant community and its dependents in Kuwait saw successive periods of comfort and security, and then hardship and unease, so too did the growth of the city survive periods of growth and stagnation. While the exact connections between economic development and urbanization are currently under review, the Kuwait situation until recent years was conveniently simple to allow the generalization that phases of economic prosperity are contemporary with phases of economic growth. Several factors bear this out:

Until 1950 natural increase of population was negligible in Kuwait. Hence all demographic growth prior to this date can be attributed to migration.

Kuwait never possessed a rural agricultural population so that "push" factors from the countryside are

largely irrelevant to Kuwait's urban growth. At most, rural-urban migration amounted to the settlement of the Badu in the city at certain phases. Most of the early twentieth century influx of population apparently took place from East Africa and Persia.

Kuwait has throughout history been the sole centre of importance in the Shaikhdom. Subcentres and villages in Kuwait were of little importance until the post-1950 period, so that the great majority of the urbanization was focused in Kuwait City.

Finally, there is documentary evidence of the historical connection between the growth of Kuwait City and the level of economic activity in the Gulf.

Overall, three distinct phases of heightened economic activity and rapid urban growth alternating with two phases of economic stagnation and slower urban growth can be distinguished.

Beginning with the establishment in Kuwait of the 'Utub in 1716, Kuwait grew rapidly, in its first 50 years of existence. A further spur to rapid urban growth was provided by the Persian occupation of Basra and the diversion of much of Basra's trade to Kuwait. The second phase saw a slackening in the pace of development when Kuwait is rarely mentioned in the contemporary Western literature. Overseas commerce and pearling provide a steady if rather vulnerable base for development throughout the period. Estimates suggest that between 1770 and 1870, the population of Kuwait roughly doubled.

With Shaikh Mubarak's accession, a phase of more active economic activity and rapid urban growth began. In the 30 years up to 1908 the population again doubled itself. Both the British and Saudi blockades severely curtailed commerce in Kuwait, but urban growth was given a boost by the migration of tribesmen to Kuwait City in the face of threats by the 'Ikhwan particularly in the 1920's. This phase of economic recession in the inter-war period, relieved only by illegitimate activities such as piracy and especially smuggling of arms and gold, finally ushered in the present period of very fast urban expansion which came in the wake of oil exporting, begun in 1946.

Up till now we have seen that Kuwait's urbanization until 1946 was in many ways typical of better known trading cities such as Damascus (al-Shams) and the cities of Iran located along the Silk Road. By 1946, it is clear that a distinctive community had evolved in Kuwait united by its commercial mores but typical of many "city states" of Europe in the later Middle Ages.

Between 1946 and the present day, Kuwait's economy has undergone a rapid transition from the traditional trading and pearling economy described above to a much more complex economic stage. Kuwait, it would seem, has passed through a kind of "Industrial Revolution" compressed into just over 20 years which brought with it standards of living and overall prosperity equal to those of Western Europe. The truth of this assumption can, however, be criticised, for while Kuwait today superficially resembles a post-Industrial Revolution country such as Great Britain, with for example, metalled roads, electricity and high standards of domestic housing and sanitation, the State almost completely lacks manufacturing industry. Certainly Kuwait has undergone an economic and social "revolution" in recent years, but it is not a revolution familiar to students of the type cast model of Western Europe.

Kuwait's contemporary internal prosperity and international significance both hang on the country's role as a major exporter of crude oil. In 1966—7, 93 per cent of the State's total income was directly derived from oil revenues and most of the rest from sources indirectly associated with the oil industry. A year earlier, the Planning Board estimated the value added by industries in Kuwait (excluding oil) to the Gross National Product at no more than 3 per cent of the total (Table 6 above). Yet in 1965, less than 4 per cent of the employment lay in the mining and quarrying (which includes oil production) category of economic activities. By contrast, over 68 per cent of those gainfully employed in that year were involved in the provision of services—electricity, gas and water, commerce, transport and communication, and other services. Hence, an apparently anomalous situation has arisen, where most of the money and foreign exchange needed to run the State is derived from a sector, which directly provides only a tiny proportion of the total employment, while most of the employment opportunities are provided in the tertiary sector, itself

making only a very small addition to the national wealth. How this situation has developed will be considered in detail below.

Detailed statistics on industry and employment are unavailable until the first Census of Population was taken in 1957. Older residents remember the period with some clarity while early Bank reports and other scattered sources provide some useful background material. From these sources, it is clear that while oil exporting began in June 1946, little tangible change overtook Kuwait's economy until the early 1950's. Ship-building remained the most notable domestic industry until quite recently, and both commerce and pearling retained their salient position in the immediate post-war years. Gold smuggling was important as a source of foreign exchange and was significant enough to result in the banning of all Kuwaiti citizens from India in 1947 [1]. However, by 1949 the Imperial Bank of Iran was able to report:

"The development of oil in the area (Kuwait) is transforming the lives of the inhabitants of these barren lands ..." [2].

Henceforward, the oil industry was financially dominant in Kuwait's economic growth. Increasing sums were spent by the Government on the construction of houses, roads, schools, hospitals, electricity and water supplies, sewerage schemes, airports, and commercial docks. This lavish disbursal of the oil revenues led to a certain amount of conspicuous private spending while providing the major impetus for Kuwait's subsequent surge forward. Growing demands for labour led to a massive influx of foreign labour which was required in quantity to meet the demands of the construction industry, and in quality to meet the growing need for technicians and skilled administrators.

Initially, almost every commodity required in Kuwait had to be imported. Gradually, the import and export business grew in significance and several personal fortunes were made in this early period. Previous trading experience left the Kuwaitis with sufficient business acumen to handle the growing import-export trade successfully. In 1951 the Imperial Bank of Iran again noted:

"The prosperity of Kuwait continues to grow ... the growing resources of the State derived from the oil royalties are being applied to the general public welfare ... The merchants of Kuwait have had a satisfactory year" [3].

Despite this commercial evolution, the fabric of Kuwait City remained virtually unaltered for some time, as the air photographs of 1951 and 1967 show (Figs. 27 and 28). In the subsequent year however, British Consultants prepared a Master Plan for Kuwait's future growth and initiated a period of construction and reconstruction which continues today. As part of the construction programme and also to facilitate the transfer of funds from the public to the private sector of the economy, the Government began to buy up land in the Old City at deliberately inflated prices. The effects of this policy on urban development, the disbursal of money amounting to over K.D. 58 million by 1957 and to ten times that amount by 1967 [4], led to an efflorescence of the consumer-orientated industries, especially retailing.

In the early 1950's, the speed of development was so fast that demand outran supply, producing inflation and a minor trade recession. By 1954, the British Bank of the Middle East reported that the "pace of development was now more in keeping with the physical possibilities" [5], and a year later that "the rise in standards of living was now more apparent" [6]. This period marks the beginning of what in other contexts has been called by Rostow the stage of "high mass consumption" [7] when Kuwait's main industry became the provision of services on a massive scale. To examine further this critical period it is of prime importance to assess the growth and organization of the oil industry which supplies the foreign exchange for all other sections of the economy.

only K.D. 15 million (K.D. 1 = £ 1 up to November 1967) had been paid directly to the Kuwait Government. However, the Kuwait Oil Company's demands for labour were steadily increasing up to 1958; since which date they have declined. The employment of well over 7,000 individuals throughout the 1950's was an important factor in the expansion of the Kuwait economy as a whole at this crucial period. At present, direct oil company employment is dwarfed by employment in

Table 8. *Oil production in million metric tons (italics) and million American barrels:*
Revenue in Kuwaiti Dinars 1946—1969

Year	Kuwait Oil Company		American Independent Oil Company		Arabian Oil Company		Total production		Revenue million K.D.
1946	*0.8*	5.9					*0.8*	5.9	
1947	*2.2*	16.2					*2.2*	16.2	
1948	*6.4*	46.5					*6.4*	46.5	5.0
1949	*12.4*	89.9					*12.4*	89.9	
1950	*17.4*	125.7					*17.4*	125.7	4.0
1951	*28.3*	204.9					*28.3*	204.9	6.0
1952	*37.8*	273.4					*37.8*	273.4	20.0
1953	*43.5*	314.6					*43.5*	314.6	59.0
1954	*48.0*	347.3	*0.4*	3.0			*48.4*	350.3	69.2
1955	*55.0*	398.5	*0.6*	4.3			*55.6*	402.8	100.0
1956	*55.2*	399.9	*0.8*	5.8			*56.0*	405.7	104.3
1957	*57.5*	416.0	*1.7*	11.6			*69.2*	427.6	110.1
1958	*70.4*	509.4	*2.1*	14.7			*72.5*	524.1	127.3
1959	*69.7*	504.8	*3.1*	21.1			*72.8*	525.9	167.3
1960	*82.1*	594.3	*3.6*	24.9			*85.7*	619.2	159.5
1961	*82.9*	600.2	*4.2*	29.3	*0.5*	3.8	*87.6*	633.3	167.0
1962	*92.4*	669.3	*5.0*	34.4	*1.6*	10.9	*99.0*	714.6	173.3
1963	*97.4*	705.5	*5.2*	35.6	*3.4*	24.1	*106.0*	765.2	190.6
1964	*107.0*	774.8	*5.1*	35.5	*4.5*	31.8	*116.6*	842.1	206.2
1965	*109.4*	791.9	*5.3*	36.5	*4.7*	33.1	*119.4*	861.5	217.6
1966	*114.7*	830.5	*4.3*	29.6	*6.6*	46.5	*125.6*	906.6	231.7
1967	*115.6*	836.7	*3.6*	24.8	*7.3*	50.9	*126.5*	912.4	241.8
1968	*122.4*	886.1	*3.1*	21.3	*8.1*	56.5	*133.6*	963.9	276.0
1969	*129.8*	940.1	*3.0*	20.9	*8.7*	61.2	*141.5*	1,022.2	295.7

Notes: 1. From 1960 onwards payments shown are for financial and not calendar years.
       2. The American Independent Oil Company's production amounts to 50 per cent of all oil produced in the Neutral Zone.
Sources: 1. Revenue from: Ministry of Oil and Finance, Kuwait, personal communication.
       2. Production from: Ministry of Finance and Oil, *The Oil of Kuwait*, Kuwait, 1965, and Institute of Petroleum, London.
       3. Early revenue payments (up to 1953) from I.B.R.D., "The Economic Development of Kuwait", p. 85, Baltimore, 1965.

After the end of the Second World War, shut-in wells were quickly placed in production and the first cargo of oil left Kuwait in June 1946. Further exploration and drilling raised production rapidly, amounting to over 200 million barrels in 1951, doubling by 1955 to over 400 million barrels, and again by 1964 to 842 million barrels (Table 8). This very rapid start to oil production is in part due to favourable physical factors in Kuwait (such as the location of the high-yielding Burqan field 8—10 km from the coast), and also to other geological factors. External factors are of prime importance in Middle East oil production and Kuwait was fortunate that as her oilfields came on stream world demand for oil was steadily rising. The Anglo-Persian dispute of 1951—4 was a cogent factor in accelerating Kuwait's offtake, just as the Suez Crisis of 1956 and the June War of 1967 both caused slight halts in the rate of production increase (Table 8).

Despite this rapid start in production revenues increased at a slightly slower pace initially. Up to 1951

other, non-producing sectors, but employment provided in the exploration and early production phases was significant in beginning the spiral of rising demands for goods and services.

In spite of early exploration and subsequent finds north of Kuwait Bay (Fig. 8), the nucleus of the oil industry remains in South Kuwait revolving around the Company town of Ahmadi and the loading terminals at Mina' al-Ahmadi and Mina' 'Abdulla. The Kuwait Oil Company (K.O.C.) was first to produce oil in Kuwait and it still retains the lion's share of production despite the addition of the American Independent Oil Company's ("Aminoil") production in 1954 and the Arabian Oil Company's production in 1961. Both these newer companies produce in the Neutral Zone, the latter from an offshore concession. A fourth company, the State-owned National Petroleum Company, was formed in 1960 to undertake exploration and production in its own right; as yet, its main task is the local retailing of oil products from K.O.C.'s refinery, augmented since 1967 by a supply of its own refinery products from Shuaiba.

None of the oil companies retain more than a handful of office staff in Kuwait City. Much of the significance of oil developments in Kuwait is based on this decentralization tendency which oil discoveries in South Kuwait have had on the pattern of population distribution. Both Ahmadi and the expanded coastal village of Fahahil are the first sizeable agglomerations of people in Kuwait's history to be located outside Kuwait City. Further industrial development at Shuaiba and a scheduled new town development called Sabahiya will both accentuate this decentralization tendency established by early oil-field exploitation. Apart from this effect on population distribution, the role of the oil industry in Kuwait has been to provide the necessary capital for the efflorescence of a variety of administrative and service industries functionally divorced from the activities of the oil sector.

Kuwait lacks traditional handicraft industries such as carpetweaving or metal-working as in several Iranian and other Arab cities. Instead, throughout its history it has concentrated on the provision of commercial and "professional" services (e. g. warehousing and pearl-broking) associated with its role as an entrepot port. It was not difficult for its merchants to replace these with newer aspects of the import-export trade without initiating any other activities such as manufacturing (Fig. 9). Kuwait's small home market and low import duties are effective barriers to large-scale industrialization.

Before the oil boom, Kuwait had a sizeable proportion of its population employed in services. Lorimer's analysis [8] indicated the presence of at least 900 retailing establishments together with an additional 250 grain warehouses. Included amongst these were several handicraft specialists (leather workers, quilt makers, blacksmiths, tinsmiths, and oil-pressers) which still persist in the old suq today. However, even by 1957 the structure of employment had changed beyond recognition from this and the immediate post-war situation. Female employment is relatively insignificant in Kuwait.

Several important points should be emphasised. Employment in the services sector occupied half the economically active population in both 1957 and 1965.

Construction is Kuwait's second major industry by employment; between 1957 and 1965, labour needs rose remarkably from just over 10,000 to almost 30,000.

Services maintained their proportional hold on employment but total employment expanded more slowly than that in construction.

Commerce (both wholesale and retail) was another fast-growing industry between Censuses.

Notably, employment in mining and quarrying slipped back, marking the phase of labour-intensive oil production consequent upon the decline of exploratory drilling in the oil industry as a whole, and the introduction of modern techniques in engineering and management.

There is a clear-cut distinction between the deployment of the indigenous Kuwaiti labour force and the immigrants—the non-Kuwaitis. Non-Kuwaitis predominate in construction and manufacturing while Kuwaitis are preponderant in the services sector. Kuwaitis have lower activity rates than the immigrants (only 35 per cent of the Kuwaiti male population was economically active in 1965 compared to 76 per cent of the non-Kuwaiti males).

By status, Kuwaitis emerge as the managers and employers while the immigrants are mostly employees; only 50 per cent of the Kuwaitis gave their status as employees in 1965 compared with 77 per cent of the non-Kuwaitis [9].

Finally, almost 20,000 Kuwaiti males in 1965 said they were "able to work but did not need to work"—a measure of the affluence of the indigenous population.

Statistics on manufacturing in Kuwait are available for 1963 and 1965. Separate censuses were taken at these dates for Government and non-Government establishments and are published in Arabic. The 1965 Censuses are more comprehensive than those in 1963; the most recent publications form the statistical base of this section.

In 1965, there were 2,325 working non-Government industrial establishments [10]. Workers employed in these establishments totalled 14,817, giving an average figure of 6.4 employees per establishment.

Overall only 5 types of establishment employed over 1,000 workers (transport equipment, non-metal products, bread making, metal products, and soft drink preparation). Together these activities were responsible for two-thirds of the total employment in manufacturing in Kuwait. The metal products division involved a variety of metal-working—mostly of products destined for the housing or construction industry while the non-metal, products division was concerned with the production of bricks, tiles, and pipes for the building trade. Clearly, what is classified as manufacturing in this context is almost a misnomer, for the manufacturing activities discussed above are very largely an extension of the service sector and would not normally be regarded as true manufacturing. With almost 4,000 people employed in the repair of motor cars forming the most numerous group in the manufacturing census, the basis of Kuwait's industrial classification breaks down. It seems that what are called manufacturing trades are little more than service industries in disguise with the emphasis placed on "repairing" rather than "manufacturing".

The industries which exist in Kuwait at present cater for the local market in the production of construction materials, the maintenance and repair of goods and vehicles, or else are connected with petroleum or gas. A few provide for simple consumer requirements—soft-drinks, flour, or tailoring—but overall, a tentative input-output table of the industrial sector indicates that at least 82 per cent of all industrial activity is in some way based on construction demand.

Conscious of this disequilibrium in the economy, the Government commissioned a British firm, Industrial and Process Engineering Consultants, to examine the possibilities of establishing new industries in Kuwait. Suggestions included plants producing tyres, car batteries, fertilizers, an oil refinery, a steel plant, a petrochemical industry, ship repair yards, an aluminium smelter, textile, cement, paper, soap, and fish canning industries [11]. Similar projects were put forward by a mission from the International Bank for Reconstruction and Development [12]. As a step towards diversification, the Government designated an area in South Kuwait near the old fishing village of Shuaiba as an industrial estate in 1961. Work began in the mid-1960's in building a combined power generating and water distillation plant with an output of 210 megawatts and 3 million gallons of water per day. In 1964, a Kuwait Chemical Fertilizer Company was formed with Government participation and began the production of ammonia, ammonium sulphate, sulphuric acid, and urea in 1966 (Fig. 10). A deep-water commercial port constructed at Shuaiba (Fig. 11) facilitates export of these products and those of the Kuwait National Petroleum Company's oil refinery which came on stream in 1967.

As part of this new drive to attract heavier manufacturing industries of the Shuaiba area of South Kuwait, the Government are investing heavily in a new town called Sabahiya between Ahmadi and Mina'al-Ahmadi which is being equipped with all services and utilities as a dormitory town for the new industrial zone. An industrial law was passed in March 1965 empowering the Government to provide a variety of incentives to industry including exemption from import duties for capital goods and raw materials, tariff protection, subsidized water and power rates, and preference in Government purchases for locally manufactured products. While Kuwait's financial resources guarantee the success of these schemes in the short term, Kuwait's basic problems remain unaltered:

"Rapid industrial development is handicapped by absence of raw materials other than crude oil and natural gas, narrow size of the home markets, high labour costs, paucity of technical, organizational and entrepreneurial skills, and the absence of motivation to join the industrial

labour force because of the open Government payroll for Kuwaitis" [13].

Despite the enormous capital investment in new industry, and estimated 42 per cent of the labour force in Kuwait relies entirely on wages and salaries paid directly by the State.

In the post-war period, Kuwait's economy was revolutionized by the development of the oil industry in South Kuwait. The revenue payments from oil exports mounted steeply and initiated a sequence of economic, demographic, and social changes which transformed the employment structure very rapidly in the early 1950's. Oil developments in South Kuwait led to a re-organization of the population distribution and a huge increase in that population's per capita wealth.

Government spending was on such a scale that two major changes took place in Kuwait. First, the Old City was expanded both vertically and horizontally in a very short space of time; inside ten years, the area covered by the city increased almost ten times. Second, a massive influx of immigrants required for all levels of unskilled and skilled employment in Kuwait modified the demographic and social attributes of the Kuwait population, and created an almost unique situation where Kuwait citizens by 1965 were in a minority in their own country.

V. Population Growth in Kuwait

1. Introduction

Parallel with the rapid post-1945 expansion of the economy of Kuwait was an equally rapid process of demographic growth. Between the first reliable estimate of Kuwait's population early this century and the most recent census of population in 1965, the number of Kuwait's inhabitants grew by 1,234 per cent (Table 9). While the numbers involved in this increase are comparatively small on a world scale, they assume much greater significance within the narrow confines of the territory of Kuwait. Further, this process of demographic growth, representing a combination of in-migration coupled with extremely high rates of natural increase, introduced into Kuwait ethnic groups quite new to eastern Arabia. The majority of the new arrivals were derived from the Arab culture area, but significant numbers of people from Africa, Asia, Europe, and the Americas are currently resident in Kuwait.

When the results of the 1965 census were published, indicating that Kuwaiti citizens were outnumbered in their own country by foreign arrivals ("non-Kuwaitis" in official parlance), the political, economic, and social implications of this situation were fully considered for the first time. A nationality law had been in force since 1948 which stipulated that Kuwaiti citizens were those individuals and their offspring who had been permanent residents of Kuwait since 1899 (Law number 2, Article 2, 1948). Citizenship could be obtained by marriage to a Kuwaiti husband or by permanent residence in Kuwait for at least 10 years. A Decree by the Amir in 1959 altered this legislation:

"Kuwaitis are basically those people who inhabited Kuwait before 1920 and have continued to reside there

until the date of publication of this law. Ancestral residence is considered complementary to that of offspring."

(Amiri Decree number 15, Article 1, 1959, translation.) Naturalization conditions became more stringent, insisting on 15 years continuous residence in Kuwait (8 for Arabs) together with a knowledge of Arabic and freedom from criminal convictions. This was further amended by Amiri Decree number 2 of 1960 when the

Table 9. *The population of Kuwait, 1904—1965*

Year	Total population	Of which Kuwaitis	Source
1904	35,000 (Excluding 13,000 Badu)	N.A.	Lorimer, J. G.: Gazetteer of the Persian Gulf, Vol. II. Calcutta 1908, p. 1051 and 1074.
1916	30,000—40,000 "possibly higher"	N.A.	Admiralty War Staff: Handbook of Arabia, Vol. II. London 1916, p. 400.
1944	70,000	N.A.	Admiralty Naval Intelligence: Iraq and the Persian Gulf. London 1944, p. 149.
1952	160,000	N.A.	Dickson, H. R. P.: Kuwait and her neighbours. London 1956, p. 40.
Feb. 1957	206,473	113,622	Census of Population, 1957, Table 1 (Arabic).
May 1961	321,621	161,909	Census of Population, 1961, Table 1 (Arabic).
April 1965	467,339	220,059	Census of Population, 1965, Table 1 (Arabic).

requisite period of residence for Arabs was extended to 10 years instead of 8. Finally, an amendment to Article 4 limited the total number of naturalizations in any one year to a maximum of 50, effectively precluding the attainment of Kuwaiti nationality by the majority of the immigrant community.

With this legislation, the Kuwaitis have successfully retained a distinction between themselves and the flood of new arrivals from a wide range of foreign countries. This differentiation is fortunate for demographic purposes but has introduced a duality into Kuwait's national life which was never significant in earlier historical periods. Before oil was discovered, Kuwait certainly consisted of a cosmopolitan blend of people from the Gulf littorals and from south Asia, coupled with an important negroid strain largely derived from slave trading with East Africa (Figs. 12 and 13). Early this century, Kuwait contained over 100 households from Najd, at least 1000 Persians, 100—200 Jews, and over 4000 Negroes, amounting to at least 16 per cent of the Shaikhdom's total population [1]. With the exception of the Negroes, two-thirds of whom were still enslaved (*Mamluk*), there are no references to any social, economic or spatial divisions of this population.

Before tracing some of the implications of the division of the Kuwait population into citizens and aliens, a closer study of the two elements of demographic growth—international immigration and natural increase—is called for. Such an analysis is necessary not only because of the repercussions of these twin factors on the population size, structure, and distribution within Kuwait, but also because very few other examples provide such a clear cut illustration of the mechanisms of migration and the lowering of mortality as does Kuwait.

2. Population Expansion by Immigration

a) War-time immigration

Between 1937 and 1947 the number of immigrants in Kuwait increased sharply, especially in the 5 year period 1942—1947. Dickson describes wartime in Kuwait as a period of difficulty for everyone (Chapter III). It seems unlikely that immigration was significant during this period, but as soon as hostilities ended and oil exports began in 1946 a flood of arrivals apparently reached Kuwait. Confirmation of this is available in the retrospective calculations based on the statistics available in the 1957 Census of Population concerning period of arrival of the non-Kuwaiti population.

Table 10 shows that up to 1947 citizens of the neighbouring Arab countries—Iraq, Iran, Oman, and Saudi Arabia—were the most numerous nationalities in Kuwait. K.O.C.'s needs were primarily for manual labour which were satisfied largely from Iran and from the Badu. Between 1937 and 1942 the Irani population, renowned for its willingness to undertake arduous manual tasks shunned by other Arabs, doubled in size. Towards the end of the periods, ending in 1947, the composition of the immigrant population showed a series of significant changes. Apparently Kuwait's riches and opportunities had become sufficiently well known to attract immigrants from beyond the countries bordering on Kuwait. Indians and Pakistanis began to arrive in growing numbers as did Arabs from the Levant (Lebanon and Syria).

Table 10. *Numbers and composition of migrants to Kuwait up to 1947*

Nationality of immigrants	Percent nationality composition				
	Before 1917	1917—1927	1927—1937	1937—1942	1942—1947
Iraqi	17	10	17	22	29
Irani	48	48	42	37	22
Omani	2	10	13	16	17
Saudi	21	20	18	12	7
Indian	1	1	1	3	5
Pakistani	1	1	3	2	5
Others	10	10	9	8	14
Total:	100	100	100	100	100
Numbers involved:	144	152	507	879	2,785

Calculated from:
Census of Population, 1957, Tables 50 a and b (Arabic).

In part this can be explained by the Oil Company's recruiting efforts in Beirut and Bombay aimed at introducing artisans and workers for more specialized tasks into Kuwait. Recruiting offices were maintained in Beirut and Bombay until the mid-1950's.

Nationality is not an infallible guide to the social characteristics of the immigrants reaching Kuwait; there are, however, obvious distinctions between the educational status of, for example, Britons and Iranis in Kuwait. With the rapid post-war increase in immigration, contrasting socio-economic groups within the same nationality grew up in Kuwait. Iranis, for example, who today form most of the manual labour force in Kuwait, also comprise a small, prosperous and influential merchant class based on a few families who are long-established residents of Kuwait. Indian traders too, who composed a small but prosperous merchant class in the Gulf ports before the last War [2], are now outnumbered by the hundreds of clerks and foremen employed by the oil companies, although the Indian community still maintains its supremacy in sections of the cloth and clothing business in Kuwait.

The oil boom, while it resulted in a sizeable increase in immigration from a greater variety of source areas than before, also attracted to Kuwait different social and economic groups from the same source area. This aspect of the immigration pattern will be considered more fully below.

b) Post-war immigration

After 1947 the immigrant population expanded rapidly: a total of 93,000 aliens were enumerated in the Census of 1957. In this decade statistics derived from the retrospective estimates are of greater reliability because of the relatively short period intervening between the dates of immigration and subsequent enumeration in 1957. Tables 11 and 12 describe the course of this immigration by nationalities in numbers and percentages.

Broadly, the statistics in these tables confirm the circumstantial evidence presented in Chapter 4 that the main phase of economic expansion and hence population immigration did not begin until the early 1950's. Since the Census was taken in February 1957, statistics do not conform to calendar years, but between 1950—51 and 1951—52 the number of migrants in Kuwait more than doubled. A further doubling took place between 1951—52 and 1953—54, followed by proportionately smaller but still sizeable increases up to 1957. Incomplete census enumeration led to a discrepancy of 28,209 between the enumerated total of non-Kuwaitis and those providing length of residence information in 1957. Hence, while the proportion of migrants arriving in any one year vis à vis other years is correct, actual numbers may be about one-third too low in every instance.

increasing numbers. The change is especially sharp after 1951 when Palestinians in particular began to arrive in increasing numbers annually (Table 11). By 1957 the five most numerous foreign groups were Iraqis (26,035), Iranis (19,919), Jordanians and Palestinians (15,173), Lebanese (6,829), Omanis (6,380), and Syrians (2,145).

Initially, the first migrants reaching Kuwait were almost all males; a retrospective calculation based on Table 11 shows that 76 per cent of the 1950 alien population were males. Even in 1957 over 80 per cent of the Irani, Iraqi, and Omani populations were males. While

Table 11. *Foreign nationals arriving in Kuwait by individual years up to 1957*

Nationals from	Before 1947	1947	1948	1949	1950	1951	1952	1953	1954	1955	1956	Total arrivals	Percent males
Iran	1611	185	300	461	625	1345	1590	2024	3348	3795	1470	16,754	96
Iraq	2476	190	371	450	513	1426	1304	1583	1999	2415	2361	15,088	77
Jordan	665	59	113	139	268	813	1827	2457	2125	3201	1476	13,143	82
Lebanon	299	6	22	45	100	319	632	912	1011	1565	1294	6,205	81
Oman	917	96	187	196	249	352	434	552	656	1141	849	5,629	93
India	744	257	215	197	77	211	462	382	397	427	380	3,749	73
Pakistan	354	106	135	87	81	149	213	247	268	363	282	2,285	75
Syria	239	6	16	27	34	98	211	227	320	491	451	2,120	86
U.K.	99	92	141	138	118	183	250	227	204	301	200	1,953	55
U.A.R.	333	2	5	6	12	76	111	197	365	484	284	1,875	50
Others	1038	17	169	144	127	255	386	545	550	762	622	4,716	—
Total	8775	1016	1674	1890	2204	5227	7420	9353	11243	14945	9669	73,517	81

Calculated from: Census of Population, 1957, Tables 50 a and b.
N.B. Statistics are for individual years from February to February since the 1957 Census was taken in that month.

During the decade 1947—1957 a radical change took place in the national composition of the migrants in Kuwait. Just after the war most of the immigrants were Iraqis, Iranis, and Omanis—unskilled labourers from countries which had long-established contacts with Kuwait. Subsequent arrival statistics (Table 12) indicate that while immigration from Iraq and Iran continued at high levels, Palestinians and Jordanians, Lebanese, Egyptians, Britons, and other Europeans were arriving in

Table 12. *National composition of the foreign born population of Kuwait up to 1957 (cumulative)*

Nationality	Percent of the non-Kuwaiti population			
	Up to 1947	Up to 1950	Up to 1953	1957 Census
Iraqi	27.9	26.1	23.8	28
Irani	18.1	19.1	21.7	21
Jordanian	7.5	7.3	13.8	16
Lebanese	3.4	2.8	5.0	7
Omani	10.3	10.4	8.6	7
Indian	8.3	10.6	7.7	4
Pakistani	4.0	5.1	4.0	3
British	1.1	3.5	3.6	2
Syrian	2.7	2.1	2.2	2
Egyptian	3.7	2.6	1.9	2
Others	12.8	10.3	7.6	8
Total	100	100	100	100

Calculated from:
1. Census of Population, 1957, Tables 48 and 50 a and b.
N.B. The slight discrepancy between the proportion of Iraqis in this Table and in Table 11 can only be explained by a reluctance of Iraqis to provide information on their date of arrival in Kuwait.

this male preponderance remained high amongst later migrants (82 per cent of the 1947—1957 residents were males—Table 11), the figure was being reduced steadily up to the 1965 Census. Females increased from 27 per cent of the non-Kuwaiti population in 1957 to 42 per cent in 1965. One of the most significant factors in this balancing trend in the sex ratios of the non-Kuwaitis is the rising proportion of people from beyond the Gulf area. As Table 11 shows, there are wide national variations in the proportions of males in the 1947—1957 residents. People from outside the Arab world display more evenly balanced sex ratios than those from within it. Egyptians are a notable exception to this rule since the U.A.R. provided Kuwait with the bulk of its female school teachers. Asians (Indians and Pakistanis) had a male bias in their populations intermediate between the Arab and the European figures. As a general statement it is apparent that a relationship exists between the level of employment obtained in Kuwait by a migrant and his willingness to bring his wife or other dependants to Kuwait. Well-paid established positions are awarded to the highly educated and technically skilled; accommodation may be provided, thus acting as a strong incentive for migrants to bring their families to Kuwait. Overall, therefore, there is a fairly clear relationship between evenly balanced sex ratios and high status employment, and vice versa, in Kuwait. Europeans tend to fall into the former category and Gulf Arabs in the latter, with Asians and educated Arabs in between.

c) Factors involved in immigration

So far, immigration has been discussed without referring to causes of this large international movement of population. Despite scanty information in the source areas of migrants in Kuwait (Census tables show only a migrant's nationality—not his place of birth or last town of residence), enough evidence is available from other sources to suggest factors relevant to immigration. The permanency or otherwise of immigration to Kuwait is an important topic which will not be discussed here but in a later section.

First, within Kuwait, there are several important "pull" factors:

1. Kuwait offers well-paid employment at all levels for people with a great variety of skills.

2. In addition, the State provides free education and a free health service, both with standards well above the average for the Middle East as a whole. These facilities are open to all non-Kuwaitis, but some schools are now closed to immigrant children because of over-crowding.

3. Kuwait's low import duties make products of foreign origin seem relatively cheap in Kuwait. Luxury goods such as radios, televisions, cameras, cosmetics, and other luxury goods, are thus readily available to the immigrant, perhaps denied them in their own country by protective tariffs.

4. The service bias of Kuwait's economy (Chapter IV) enables a migrant with the minimum of technical training to obtain a job in, for example, Government administration or its ancillaries.

Similarly, there are several well-defined "push" factors in the home areas of the immigrants:

1. Overall, levels of prosperity in the Middle East are below those prevailing in Kuwait. This is particularly true of neighbouring areas such as southern Iran and southern Iraq.

2. A study of national Censuses shows (Chapter I) that there is a regional drift of population from rural to urban areas. Kuwait City plays a part in this trend, not only in the settlement of the Badu in Kuwait, but also in the movement of rural dwellers from other states to Kuwait City. Detailed studies are lacking, but the most recent Jordanian Census (taken in 1961 and published in 1964) shows that 50 per cent of the Jordanian population living abroad are resident in Kuwait. Of these 31,739 people, 21,628 were from the Nablus area [3]. This district is undoubtedly a rural area. Anthropologists working in Iran suggest that the bulk of the Persian immigrants in Kuwait are also from rural source areas. Azeez also has some important information on migration within Iraq [4].

3. Wider political forces are also at work in causing large-scale movements of people to Kuwait. Most significant among these is the establishment of Israel in 1948 and its subsequent expansion over most of what was Palestine. Obviously this expansion is the most important factor in causing the increase in the number of Palestinian and Jordanian arrivals in Kuwait after 1950 (shown in Table 11). Indian independence in 1948, and the war with Pakistan, is also an important factor in increasing Asian immigration to Kuwait. Major changes of Government and several coups d'état in the neighbouring countries were also instrumental in hastening immigration into Kuwait.

4. A further factor in the determination of the source area of the migrants has been the growing maturity of the Kuwait economy as a whole. Initially, manual labourers were in most demand; now demands are higher for the literate and technically skilled. Such a change in emphasis may be in part responsible for the growing numbers of Jordanians (16 per cent illiterate in 1965), Lebanese (17 per cent illiterate), and Syrians (26 per cent illiterate) arriving in Kuwait compared with Iranis (81 per cent illiterate) and Iraqis (67 per cent illiterate).

d) The Alien population 1957—1965

Demographically this period is well documented with three Censuses in a space of eight years. The Censuses indicate that the non-Kuwaiti population increased by more than two and a half times between 1957 and 1965. Comprising 45 per cent of Kuwait's population in 1957, immigrants grew to form 53 per cent of the total in 1965. Assuming even enumeration at all three Censuses, the rate of immigration was slightly slower in 1961—1965 than between 1957 and 1961.

Besides this overall increase of the immigrant population the period was distinguished by several significant trends in the character of the alien population.

1. Between 1957 and 1965 the numbers of women arriving in Kuwait rose sharply. In 1957 there were only 274 non-Kuwaiti women per 1,000 non-Kuwaiti males. By 1961 the figure had risen to 374, and by 1965 there were 423 females per thousand males.

2. Associated with this increase in non-Kuwaiti women was an increase in the number of young children. By 1965 14.6 per cent of the non-Kuwaiti population were under the age of 5, compared with less than 7 per cent in 1957.

3. Over the same period the proportion of male illiterates declined from 42.9 per cent to 32.7 per cent in 1965, pointing the way towards a growing sophistication in the migrant stream. From being an almost exclusively male manual labour force up to 1957, the non-Kuwaiti population by 1965 had taken on the appearance of a settled population accompanied by its usual dependants— females, the very young, and the very old.

These trends are observed for the non-Kuwaiti population as a whole. However, the immigrant population is by no means a homogenous group. In order to clarify the important demographic changes overtaking this population between 1957 and 1965 national groups have been distinguished, and their age and sex structures represented in a series of pyramids (Text Fig. 3) (see back of Map 1). Unfortunately, the 1961 Census presents statistics on age and sex in such a form as to make comparison with the other Censuses almost impossible; as a result, only the Censuses of 1957 and 1965 are used in this comparison.

aa) Sex ratios

Broadly, the predominance of males declined between 1957 and 1965 for most nationalities. Possible exceptions include the Iranis and the Omanis. Both these groups are highly mobile (see below), making frequent return trips to their home area; the latter are known for their pattern of migratory hawking in the Gulf ports.

Four major classes of sex ratios are recognisable in the population pyramids.

1. The first class includes those national groups with over 70 per cent male populations in 1965. Groups with such high proportions of males are the Iranis (94 per cent), the Omanis (86 per cent), the Syrians (72 per cent), and people from the Trucial States and Qatar.

2. Indians and Pakistanis, the main representatives of Asian migrants in Kuwait, had closely similar sex ratios;

in 1965, 66 per cent and 65 per cent of their respective populations were males.

3. In the third group were included most of the migrants from the Arab area as a whole, with between 61 and 64 per cent males in their population. These nationalities—Jordanians, Saudis, Lebanese, and Iraqis—all increased their proportion of females between 1957 and 1965.

4. The last group with almost even proportions of males and females consisted of Europeans and Britons, Americans, and, surprisingly, Egyptians, all with less than 55 per cent males in their population. While the age-sex pyramid for Britons balances by age group, for the Egyptian pyramid the peaks of the male and female populations are separated by approximately ten years. As stated above, the U.A.R. provides Kuwait with many of its female teachers, many of them single girls, which is why overall sex ratios for Egyptians are almost evenly balanced.

bb) Age structure

Almost all the age-sex pyramids have a "waisted" appearance because of the lack of adolescents of both sexes in the foreign-born populations in Kuwait. In addition there are very small proportions of the aged. Both these groups have good reasons for remaining in the home country—the first for schooling, and the latter because they have passed the active age range.

Most of the nationalities have a fairly proportion of young children, with the exception of the Iranis, the Omanis, and people from the other Gulf states. These three groups are the most masculine populations in Kuwait, as well as paying frequent visits home (see below). Despite the high proportion of males in the active age group (20—45), dependency ratios have dropped sharply since 1957. In 1957 adults aged 15—60 outnumbered the very young and the very old by almost 5 : 1; by 1965 the ratio had dropped to 2.4 : 1. Text Fig. 3 illustrates the demographic variety of the immigrant population in Kuwait. A further aspect of immigration to Kuwait which has been mentioned above is its permanency or otherwise. Fortunately, statistics are available to check this variable by nationality.

e) Permanency of migration to Kuwait

To estimate the length of time spent in Kuwait by migrants of various nationalities, two sets of statistics are available.

1. Using arrival and departure figures it is possible to relate the frequency with which a national group crosses Kuwait's frontiers to the total enumerated population of that group resident in Kuwait.

2. Alternatively, the period of residence tables can be compared in the Censuses of 1957 and 1965 and related to nationalities.

Both methods are used and discussed below. Population stability measured by the first method is compared by a "Stability Index" below, which is the 1965 nationality population expressed as a percentage of the sum of arrival and departure figures for 1965.

Such a table has two obvious failings; it assumes 1965 was a "typical" year for arrivals and departures which a study of other years suggests it was. In addition, it fails to consider illegal immigration and emigration.

A high index indicates that the members of a national group rarely crossed Kuwait's frontiers and a low index

Table 13. *Population stability in 1965 measured by relating arrivals and departures to numbers resident in Kuwait*

Nationality	Stability index	Nationality	Stability index
Iranis	119	Jordanians and Palestinians	47
Pakistanis	78	Egyptians	45
Kuwaitis	78	Syrians	38
Indians	67	Gulf States	30
Omanis	63	Britons	15
Other Arabs [a]	49	Iraqis	9
Lebanese	47		

Average of 19 National Groups = 41.
[a] Arabs from Yemen, Sudan, Aden and North Africa.
Calculated from: Monthly Statistical Bulletins, 1965. Census of Population, 1965, Table 2.

the reverse. Apparently Iranis are the most stable group in Kuwait which other evidence strongly contradicts (see below). Iranis cross into Kuwait by sea, entering illegally; 14,400 Iranis were deported from Kuwait for this reason in 1965 (Statistical Abstract, 1965, Table 53), ten times as many as any other nationality. Of 34,875 illegal entrants in the labour force of 1963, 30,542 were Iranian [5]. Omanis may well be in the same category as Iranis, but otherwise the table is a fair description of migrant turnover in Kuwait. Iraqis cross Kuwait's borders very frequently because of their proximity to home. Britons are also frequent crossers of Kuwait's borders, probably because annual leave is included in most contracts which are rarely awarded for more than three years at a time. Asians, by contrast, are amongst the most stable residents of Kuwait. Distance to home and the structure of their age and sex pyramids are both probable explanations.

For the second method, period of residence tables for 1957 and 1965 are compared for the intervening period 1957—1965 at three dates. In 1950, 1953, and 1957, the population of the various nationality groups provided in both the 1957 and 1965 Censuses is contrasted by expressing the deviation between the two Census as a percentage error. The results of these calculations appear as Table 14.

In effect the Table measures those people enumerated in 1957 and again in 1965 who said they were present in Kuwait in 1950, 1953, and 1957. Some errors are inevitable, but the Table broadly confirms some points about migrant stability mentioned above. Those nationalities at the top end of Table 14 are believed to stay in Kuwait longest, and those at the bottom are more temporary migrants. Europeans, Britons, Americans, and Iranis all appear to be the least permanent residents—the former because of their short-term contract method of employment and the Iranis because of their proximity to home. Saudis, Indians, Omanis, and Jordanians seem to represent the more permanent element in Kuwait's alien population, appearing as they do at the top of Table 14.

The two Tables do not directly correspond because of enumeration errors and because of factors such as natural increase. Saudis, Jordanians, and Indians probably overestimated their length of stay in Kuwait in 1965. Kuwaitis themselves appear to have only a medium stability because of their propensity to travel and also because of a high rate of natural increase between 1957 and 1965.

Table 14. *A comparison of length of residence tables in 1957 and 1965*

Nationality	Average error 1957—1965	1950 Population % Error 1957—1965	1953 Population % Error 1957—1965	1957 Population % Error 1957—1965
Saudi	96	81	88	119
Indian	88	79	79	104
Omani	75	90	63	73
Jordanian	68	22	77	105
Syrian	63	38	68	83
Iraqi	61	62	59	62
Pakistani	55	72	83	110
Lebanese	52	21	65	70
Amviates	52	58	57	42
U.S.A.	37	60	17	35
British	36	28	39	42
Irani	32	35	32	30
Egyptian	32	18	32	46
Other Europeans	14	8	14	21
Kuwaitis	61	56	59	69

Calculated from: Census of Population, 1957, Tables 50a and b. Census of Population, 1965, Table 24.

Nevertheless, the Tables are complementary to a study of age-sex distribution amongst the migrants since both the residence tables and the population pyramids illustrate the variety and complexity of the factors involved in determining the origin and characteristics of the migrants reaching Kuwait.

f) Arrivals after 1965

Arrival and departure statistics are published monthly by nationalities; statistics are available for three years after the 1965 Census and are presented in Table 15.

Clearly, immigration is continuing at a high rate. Amongst the 10 most numerous foreign-born groups in Kuwait in 1965, Iranis alone decreased in numbers between 1965 and 1968. The sizeable influx of northern area Arabs in 1967 is directly related to the Arab-Israeli June war of that year. Notably, Jordanian and Palestinian immigration leapt to 34,300 in 1967, followed

by slightly smaller increases from nations closely associated with the war zone—Syrians, Lebanese, and Egyptians. Unexpectedly, Iraqi immigration also leapt upwards, perhaps because of a more general feeling of insecurity in the Arab world as a whole. Iranis presumably returned home because of this same reason.

Between April 1965 and January 1968 the non-Kuwaiti population almost doubled. Jordanians remained the most numerous (27 per cent of the non-Kuwaitis), followed by Iraqis, Syrians, and Egyptians. The June War of 1967 apparently brought about a considerable change in the composition of the immigrant population in Kuwait. Assuming that the age and sex characteristics of the new arrivals paralleled those of their countrymen enumerated in 1965, the proportion of single male immigrants will have decreased and been replaced by a more balanced family structure. Natural increase amongst immigrants is thus likely to be a much more important consideration in the future than previously thought; attention will be directed to this point in the following section.

Statistics after 1968 are not as yet available in a complete form, but from a study of statistics for the first six months of 1968 it is apparent that return migration on a large scale is not taking place. In fact, between January and June 1968 the non-Kuwaiti population increased by a further 18,200 through net immigration.

So far we have reviewed the sources of information available on immigration since the inter-war period, and have traced the increase of the non-Kuwaiti population by direct and indirect methods. While the major increase in the Kuwait population has been by immigration, natural increase by Kuwaitis and non-Kuwaitis alike is assuming growing significance in the overall population growth of the State.

3. Population Expansion by Natural Increase

Introduction

Natural increase occurs in a "closed" population when the number of births exceeds the total number of deaths over a period of years. A "closed" population implies that in- or out-migration is excluded from the

Table 15. *Net population movements by nationality from 1965 onwards*

Nationality	1965 Census population	1965 Net movements	1966 Net movements	1967 Net movements	January 1968 Population totals	Percent 1968 Total
Jordanian	77,712	+6,847	+847	+34,288	119,694	27.1
Irani	30,790	—7,851	—1,834	—12,028	12,488	2.8
Iraqi	25,897	+5,091	+14,849	+28,956	74,793	16.9
Lebanese	20,877	—179	—915	+7,115	26,898	6.1
Omani	19,520	+255	—928	+6,182	25,029	5.7
Syrian	16,849	+1,547	+6,956	+16,471	41,823	9.5
Pakistani	11,735	+191	—942	+2,673	13,657	3.1
Indian	11,699	+8	+164	+2,771	14,642	3.3
Egyptian	11,021	+1,129	+4,066	+4,709	20,925	4.7
British	2,837	+125	+125	+630	3,717	0.8
Total:	229,937	+7,163	23,596	+91,767	351,463	79.5
All non-Kuwaitis:	247,280	+8,423	+29,790	+156,478	441,971	100

Compiled from: Census of Population, 1965, Table 2. Monthly Statistical Bulletins, 1965—1967, Tables 2 and 3.
N.B. The 1965 Census was taken in April 1965: net movements are shown only for the succeeding 8 months of that year.

calculation; when both natural increase and migration are taken into account any increase in numbers is referred to as "population growth".

Immigration, as the previous section has shown, is of prime significance in the overall growth of Kuwait's population. However, the indigenous population has also been expanding in the post-War period, mainly by natural increase. Migration and natural increase are significant in the population growth of the Kuwaiti and the non-Kuwaiti populations as a whole, although migration is of almost negligible importance in the growth of the former population. Sources of information are scattered up to the late 1950's, but sufficient statistics are available for the post-1945 period to describe the course of population growth, particularly natural increase and its causal factors.

a) Vital statistics in Kuwait

Registration of births, deaths, and infectious diseases was made compulsory in 1952, but until 1959 when a Statistical Section was established in the Ministry of Public Health records were far from accurate. Annual reports published by the Section provide an increasing variety of statistics relevant to the vital events but, as yet, the published tables are not completely comprehensive. Medical facilities have grown enormously in Kuwait recently (see below), which has had the fortunate effect of encouraging most mothers to give birth in a hospital rather than at home. There the birth is automatically registered and subsequent mortalities carefully recorded. Deaths are not so fully documented since only a small proportion of total deaths occur in hospital. Again, the speed with which burial follows death in the Arab World is an important factor in the omission of the registration of death.

In addition, deaths of non-Kuwaitis or even Kuwaitis may occur in other countries; as yet, no attempt is made to collate domestic and foreign deaths for the two populations, so that published mortality rates have only a limited validity. On the other hand, Kuwait's high quality and largely free medical services are an encouragement to non-Kuwaitis living outside Kuwait to bring their wives to the State and to have their families within its frontiers. As a result, birth rates may well be above what would normally be expected.

Besides the direct records of the vital events by the State Health Service, two further sources of information are available. First, the Kuwait Oil Company runs its own health service for its employees and dependants at Ahmadi, in all covering a population of over 20,000. Detailed and accurate records are kept for this population.

Second, the Municipality of Kuwait maintains a burial book giving the numbers of graves dug and filled each year. Some of these records are published in the Municipality's annual reports, but together with unpublished data they provide a means of checking other mortality statistics.

Finally, the three population Censuses of 1957, 1961, and 1965 can be used in a variety of ways, if not to produce a complete life table, at least to check to some extent both mortality and natality statistics.

b) Natality, mortality and the population cycle

Kuwait's recent demographic history closely parallels the experience of many countries of the developing world. A pattern of declining mortality rates and almost stable birth rates has been recognized during the nineteenth century in England and Wales [6] as well as in several contemporary situations [7]. The transition from high fluctuating birth and death rates to the stage of low fluctuating rates reached in advanced societies today has been called the "population cycle". Statistics available for Kuwait (Table 17) indicate that the stage of "early expanding" population growth has been reached in which mortality has declined to low levels (less than 20 per thousand) while the birth rate remains very high (over 35 per thousand). As a result, a sizeable natural increase in the total population is occurring.

One of the most important elements in the lowering of the death rate of a population is the provision of modern medical care and facilities. Infant mortality, that is, deaths occurring amongst children within one year of birth, is most susceptible to rapid lowering by the application of medical care. The extension of life by the postponement of death until old age is a much slower process [8]. Most of the excess of births over deaths in developing countries today arises because of the increased numbers of young children "saved" from death rather than because of increased fertility. However, since births and deaths are always associated in some way with age, the demographic characteristics of a population, as well as the scope and quality of the medical care available to it will both be potent forces in the determination of the vital rates and hence the scale of population increase.

c) Health facilities in Kuwait

aa) Growth

Early this century, modern medical facilities were unknown in Kuwait although a traditional medical code was in use (see below). An American doctor, together with several semi-trained nurses, arrived in Kuwait in 1909 as part of the Arabian American Mission of the Dutch Reformed Church and began the establishment of modern medical facilities on a small scale. By 1911, a small men's hospital had been opened, followed 8 years later by a one-storey hospital for women [9]. Finally, a larger hospital was opened in 1939—the Olcott Memorial Hospital—with 34 beds, one or two doctors, 4 nurses, and 10 illiterate women helpers and servants [10]. While the Kuwait Oil Company provided a tented hospital for its employees in the pre-war period, and a more permanent clinic at Maqwa subsequently, it was not until 1954 that it opened a small clinic for wives and children of its Arab employees at Ahmadi [11].

Formed in 1936, the Ministry of Health in Kuwait took over responsibilities for its citizens in the post-war period, beginning with a 100-bed hospital opened in 1949. Inside 10 years the number of beds had increased to 582 while a variety of specialist clinics have been added, notably those dealing with maternity disorders and women's diseases. Numbers of doctors employed and hospital beds available grew rapidly (Table 16) as the State Health Scheme expanded the scope and the quality of its service. Clinics, hospitals, and special care clinics were opened with bewildering speed throughout the 1950's; a tuberculosis sanatorium was opened in 1952, a new mental hospital in 1955, a difficult birth unit in 1953, an infectious diseases hospital was opened in 1957, and the first of many motherhood and infant care centres was opened in 1955. A new hospital—the Sabah—supplemented the facilities of the older Amiri hospital in 1962 and special orthopaedic clinics followed. Parallel developments took place in the private medical field, with new hospitals, clinics, (and specialists to staff them), proliferating up to the present day.

Table 16. *The expansion of medical facilities in Kuwait*

Date	Estimated total population	Total number of Doctors [a]	People per Doctor	Hospital Beds provided by the State	Private Hospital Beds
1949		45		100	34
1953	170,000?	46	3696	611	N.A.
1957	206,000	145	1421	1,322	N.A.
1962	402,000	462	870	2,600	N.A.
1966	506,000	606	835	3,002	384

[a] Includes doctors working privately outside the State Health system.

Constructed from: Statistical Abstracts, 1964—1968, "Health Statistics". Ministry of Health, Statistical Reports, 1958—1966. Ministry of Health, personal communication.

bb) Effect

Since our interest at present is not in the development of the facilities of the Kuwait Health Service *per se*, but rather in their demographic effects, further details of the facilities themselves are unnecessary. The foregoing account suffices in that it places the beginning of the decline in mortality—especially infant mortality—at about the time of the Second World War and possibly later. While Dr. Calverley's early mission hospital undoubtedly relieved individual suffering, it is doubtful whether it had any large-scale effect on the health of the population at large. From Lorimer's evidence it is clear that health standards were low in the early years of this century and the risk of dying very high. He writes, for example, that Kuwait possessed open cesspools for sewage in the city centre [12], and that cholera, bubonic plague, smallpox and malaria were all frequent and fatal visitors to the northern Gulf area [13]. Epidemics of cholera killed 7,000 in Bahrain in 1893 and a further 1,200 in 1904 [14]. Half the inhabitants of Najaf in Iraq were killed by bubonic plague in 1881 and 400 died at Basra in 1892 [15]. Smallpox in 1900 killed 500 in Sharja alone [16]. Lorimer described the inhabitants of Kuwait as "long faced, sickly, and inclined to lankiness" [17], possibly because of their poor health.

Small inroads were made into this pool of ill-health by the Mission Hospital and K.O.C. before 1949. Major alterations in the pattern of community health only began after the Amiri State Hospital with 100 beds was opened in 1949. Thenceforward, a steady improvement in health was recorded until today medical standards as a whole compare favourably with those of an evolved European country.

On the basis of this evolution we can recognize several medico-demographic stages constructed from both direct statistical and indirect circumstantial evidence:

Table 17. *Stages in the reduction of the mortality rates in Kuwait*

Stage	Date	General mortality rate per thousand	Infant mortality rate per thousand	General characteristics of the period
1	Before 1909	25+	120+	Pre-medical stage with almost no natural increase
2	1909—1946	20—25	100—120	Mission Hospital; some local effect on health. Small natural increase.
3	1946—1950	17—23	80—100	Transition stage; State Health Scheme augments K.O.C. and Mission Hospital.
4	1950—1955	12—18	50—70	Over 1,000 hospital beds available; maternity care begins in earnest.
5	1955—1969	10—15	35—45	Comprehensive health service emerges; less than 1,000 persons per doctor. Infant mortality equivalent to rates in U.K. in late 1940's.

Sources: 1. Pre-1955 data from a comparative study of other developing nations, e.g. U.N. Demographic Yearbook, 1966. Tables 14 and 17.
2. Post-1955 data from Ministry of Health, Annual Reports, 1957—1969.

Particularly clear is the reduction of infant mortality rates in the post-war period; in just over 20 years infant death rates were halved from 80—100 to 40 per thousand, a change which took over 40 years in England and Wales. At present, total mortality is at a rate slightly below that of the U.K. because of the youth of the total Kuwait population (38 per cent under age 15). This element—the age-structure of the population—merits attention as the second major factor affecting population increase.

d) Age structure of the Kuwait population

Age-sex pyramids for the Kuwaiti and non-Kuwaiti populations at the three Census dates appear as Text Fig. 4, back of Map 1. Clearly, the two populations have quite distinct demographic characteristics. The Kuwaiti population pyramid has a roughly triangular shape indicating that it is expanding rapidly by natural increase while the non-Kuwaiti population pyramid is of a quite exceptional form. For this reason the Kuwaiti and non-Kuwaiti populations will be separately treated.

aa) The Kuwaitis

With almost half of their number under the age of 15, Kuwaitis were among the "youngest" populations in the world—if not the youngest. As Table 18 shows, the proportion in this youngest age group has been growing steadily since 1957; by 1965 the median age of the Kuwaiti population had fallen to 14.

So marked was this increase in the proportion of children that by 1965 less than half the total Kuwaiti population was in the active age groups (15—19). Dependency ratios (the ratio between the active and inactive sections of the population) are thus extremely high and show no signs of decreasing (see natality statistics below). In general, Table 18 substantiates the evidence of the population pyramids (Text Fig. 4, back of Map 1) indicating that the Kuwaiti population is undergoing a period of very rapid growth by high natural increase. The magnitude of this growth will be reviewed below.

Table 18. *Percentage age distribution of the Kuwaiti and non-Kuwaiti populations in 1957, 1961, and 1965*

Year	0—14		15—39		40—59		60 and over	
	Kuwaitis	Non-Kuwaitis	Kuwaitis	Non-Kuwaitis	Kuwaitis	Non-Kuwaitis	Kuwaitis	Non-Kuwaitis
1957	41.5	15.2	38.8	63.8	12.6	12.6	2.5	1.5
1961	43.4	23.6	32.9	63.1	11.1	10.8	5.4	1.5
1965	49.0	28.2	35.5	60.0	10.6	10.4	4.8	1.3

Calculated from: Censuses of Population, 1965, Table 9; 1961, Table 6; 1957, Table 3.
N. B. Totals may not equal 100 since some respondents did not state their age.

bb) The Non-Kuwaitis

The age-group 15—59 has contained over 70 per cent of the total alien population in Kuwait at all three census dates (Table 18). Between 1957 and 1965, however, two significant related changes overtook the non-Kuwaitis. First, the numbers of women in the child-bearing age groups (15—44) almost quadrupled compared with a doubling of the numbers of Kuwaiti women in this category (Table 19). Thus the second change arose in the composition of the immigrant population—the proportion of children in the 0—14 age group doubled between 1957 and 1965. As other evidence suggests, the immigrant population which was initially composed mainly of young adult males searching for work in Kuwait is now taking on a more settled appearance because of the arrival of the dependants of the labour force as well as the workers themselves. High child-women ratios (Table 19) confirm the thesis put forward above that Kuwait's free medical and educational facilities are a powerful attraction to potential migrants.

Table 19. *The relationship of children under 5 to the numbers of women in the child-bearing age group (15—44) in Kuwait*

Year	Children under 5		Women 15—44		Child-women ratio	
	Kuwaitis	Non-Kuwaitis	Kuwaitis	Non-Kuwaitis	Kuwaitis	Non-Kuwaitis
1957	18,703	5,826	20,119	8,903	930	65
1961	27,294	18,491	26,813	21,044	1,018	88
1965	44,158	36,216	42,172	35,723	1,047	1,014

Calculated from: Censuses of Population; 1957, Table 3; 1961, Table 6; 1965, Table 9.

The child-women ratio is calculated by dividing the number of children under 5 by the number of women 15—44 and multiplying by 1,000. It is a crude measure of fertility since it relates the number of survivors of 5 years of births to the average number of women capable of producing these infants. Clearly, one of the factors responsible for the growing proportion of youngsters in the non-Kuwaiti population is its high

Table 20. *Child-women ratios[a] in Kuwait with some international comparisons*

Country	Date	Ratio	Country	Date	Ratio
Kuwait	1961	900	Kuwait	1965	980
Libya	1964	806	India	1961	659
Morocco	1960	845	Pakistan	1961	832
Tunisia	1956	716	Iran	1956	790
U.A.R.	1960	702	Iraq	1957	936
Canada	1961	535	Turkey	1950	700
U.S.A.	1960	488	Denmark	1960	338
Brazil	1960	667	U.K.	1961	336
Venezuela	1961	804	U.S.S.R.	1959	695

[a] Calculated as a ratio of the population under 5 years of age per 1,000 female population aged 15—49.
Source: U.N. Demographic Yearbook, 1965, Table 8.

level of fertility—at present, with the Kuwaitis, one of the highest in the world (Table 20).

Fertility will be dealt with more thoroughly below; but the age-sex structure of a population is related to fertility. For these reasons it is of great importance to examine the differences and changes in the age-sex structures of the various national groups in Kuwait. As the age-sex pyramids show (Text Fig. 4 back of Map 1) the non-Kuwaiti population of Kuwait is far from homogeneous.

In the Middle East, young adult males are the most mobile section of the population compared to the reverse situation in Britain today. One would expect to find the age-sex pyramids of the national groups corresponding closely with that shown for the Jordanians in 1957 (Text Fig. 4, back of Map 1). Numerous factors are responsible for the departure of individual national groups from this expected norm, but first, sets of broadly similar pyramids can be recognized.

1. Several nationalities display a strong male bias in the active age-groups together with a small number of dependants (the very young and the very old) and an even smaller number of teenagers. Included in this group are the Jordanians, the Lebanese, the Indians, the Pakistanis, and the Syrians.

2. A more extreme version of the first category is displayed in the pyramids for the Iranis, the Omanis, and peoples from the other Gulf States. These migrants are almost exclusively male and are probably seasonal visitors since they have virtually no dependants in Kuwait.

3. Those with a more balanced distribution—roughly equal sex ratios throughout and a more "normal" proportion of dependants—include the British and the Iraqis.

4. Two exceptional groups are the Egyptians and the Saudis. The former are older than most migrants in Kuwait and have a surplus of females in the 20—30 age-group. Saudis, by contrast, display an irregular distri-

bution probably because of wrong age recording. Many Saudis are Badu who may have wrong concepts of their true age because of lack of documentation.

e) Natality and fertility

Recently, birth rates have shown a steady upward trend in Kuwait. As the age-sex pyramids show (Text Fig. 4), Kuwaitis can be classed as a clearly "progressive" population while the factors affecting the expansion of the non-Kuwaiti population are much more complex.

aa) Kuwaitis

While the total number of births to Kuwaiti women has more than tripled between 1958 and 1966 (Table 21), both the crude birth rate (births as a proportion of the total population) and the general fertility ratio (total

young, with a well-paid job at the end of his or her education.

Birth control is growing in popularity and devices are both on open sale and available free through the health service in Kuwait. As yet, its effects seem limited.

bb) Non-Kuwaitis

As Table 22 shows, the crude birth rate for non-Kuwaitis is still substantially below that for Kuwaitis. Nevertheless, non-Kuwaitis are reproducing rapidly as sex ratios gradually even out (there were 274 females per 1,000 males in 1957 amongst the alien population; by 1965 the figure had risen to 423), and especially as the number of women in the child-bearing age-group grows rapidly (Table 18). Between 1957 and 1965 the proportion of married non-Kuwaitis in the alien population over the age of 15 rose from 39.7 per cent to 43.5 per cent.

Table 21. *Births and birth rates: Kuwaitis only*

Date	Total population	Total Kuwaiti births	Crude birth rate per thousand	General fertility ratio per thousand
1958	123,900	4,658	37.5	231?
1959	135,000	5,675	42.0	
1960	147,000	6,842	46.5	
1961	161,909	6,911	42.7	258
1962	175,200	7,921	45.2	
1963	189,500	9,261	48.9	
1964	204,900	10,014	48.9	
1965	220,100	11,291	51.3	268
1966	240,000	14,057	58.6	

Table 22. *Births and birth rates: non-Kuwaitis only*

Date	Total population	Total non-Kuwaiti births	Crude birth rate per thousand	General fertility ratio per thousand
1958	105,479	2,223	21.1	250?
1959	119,824	3,348	27.9	
1960	136,120	4,774	35.1	
1961	159,712	6,031	37.8	287
1962	178,558	7,283	40.8	
1963	199,628	8,459	42.4	
1964	223,184	9,414	42.2	
1965	247,280	9,764	39.4	273
1966	285,493	11,271	39.5	

N. B. Statistics represented for Census years (1957, 1961 and 1965) are more reliable than for the intervening years.
Calculated from:
1. Census of Population; 1957, Table 3; 1961, Table 6; 1965, Table 9.
2. Ministry of Health, Annual Reports, 1958—1962.
3. Statistical Abstract, 1967, Table 15.

N. B. Statistics presented for Census years (1957, 1961, 1965) are more reliable than for the intervening years.
Calculated from:
1. Censuses of Population; 1957, Table 3; 1961, Table 6; 1965, Table 9.
2. Ministry of Health Annual Reports, 1958—1962.
3. Statistical Abstract, 1967, Table 15.

births expressed as a proportion of women aged 15—44) have also risen over the period. In other words, proportionately more children are being born to the same number of Kuwaiti women in the child-bearing age-group than in all the earlier years.

Several factors explain this propensity to have more children among Kuwaitis:

1. Health facilities have improved greatly in recent years (see above). More young babies are kept alive than ever before and mothers run a much reduced risk during childbirth. Registration of births is also being steadily improved.

2. Rising affluence has meant that another child brings no financial problems—rather, it brings prestige to the family.

3. More young women are in the child-bearing age group than before; this will continue because of the "progressive" form of the population pyramid.

4. Family size is growing steadily and marriage occurs at a young age. 379 marriages of a total 1,820 in 1966 occurred when the bride was under the age of 19 (Monthly Statistical Bulletin, January 1967, Table 8).

5. Finally, every Kuwaiti knows that his offspring are guaranteed a free and comprehensive education while

While the crude birth rate amongst non-Kuwaitis is lower than that for Kuwaitis, it is notable that the former's general fertility ratio has been higher since 1957 (cf. Tables 21 and 22). Non-Kuwaitis have no stability of residence or long-term career prospects in Kuwait since every year both work and residence permits must be renewed. Despite this, non-Kuwaitis have free access to Kuwait's health, welfare, and educational services. It seems plausible that non-Kuwaitis use these facilities to the full, thus explaining the higher fertility amongst their women of child-bearing age.

Unfortunately births are not recorded by specific nationalities for the non-Kuwaitis in Kuwait. By assuming that all children under one in the 1965 Census have been born in Kuwait, we can obtain some indication of how fertility varies amongst the nationalities by relating these numbers to the total population of a national group. To allow for mortality in the year preceding the Census, the numbers of the under-ones are divided by the co-efficient 0.9. The results of this calculation appear as Table 23 and confirm broadly what was noted above. Calculated "crude birth rates" are highest for those populations with the most balanced age-sex pyramids; that is; Lebanese, Jordanians, Pakistanis, Indians, and

Syrians. By comparison the Irani and Omani groups display the lowest "crude birth rates".

Errors are inevitable by this method: overall, 9,764 non-Kuwaitis were born in 1965 and 9,414 the year before, compared with only 8,417 enumerated in the 1965 Census as under age one. But the calculations produce a crude birth rate of 37.8 per thousand for all non-Kuwaitis compared with an observed figure of 39 per thousand (Table 22). Hence, the relative positions of nationalities in Table 23 are probably correct, even though absolute rates are probably low.

Table 23. *Calculated "crude birth rates" for various nationality groups in 1965*

Nationality	Children under 1 year	Divided by 0.9	Total population	"Crude birth rate" per thousand
Lebanese	1,136	1262	20,877	60.4
Jordanian	3,808	4231	77,712	54.4
Pakistani	484	538	11,735	45.8
Indian	411	457	11,679	39.1
Syrian	570	633	16,849	37.6
Saudi	153	170	4,632	36.7
Egyptian	352	391	11,021	35.5
British	82	91	2,837	32.1
Iraqi	694	771	25,897	29.8
Omani	381	423	19,520	21.7
Irani	202	224	30,790	7.3

Calculated from: Census of Population, 1965, Table 23.

f) Mortality

Deaths in Kuwait are less comprehensively registered than births; of 3,603 people buried by the Municipality in the year April 1966—April 1967, only 78 had the necessary certificates [18]. To correct this under-estimation in the official statistics, recourse was made to the Municipal Burial Books and graveyard records to provide a more realistic mortality rate for both Kuwaitis and non-Kuwaitis. Graves were counted in Kuwait's main graveyards in order to check the Municipality's statistics.

Table 24. *Mortality statistics for Kuwait 1958—1966*

Year	Total deaths recorded by Ministry of Health	Crude death rate per thousand	Deaths recorded in municipal records	Corrected crude death rate	Crude death rate for Kuwaitis
1958	769	3.3			
1959	892	3.5			
1960	1,235	4.4			
1961	2,504	7.8			
1962	2,180	6.2			
1963	2,139	5.5			
1964	2,618	6.1			
1965	2,468	5.3	3,409	7.3	9.8
1966	2,813	5.5	3,510	6.8	9.5

Calculated from:
1. Total deaths from Ministry of Health, Annual Reports, 1958—1964; and Statistical Abstract 1967, Table 16.
2. Other deaths recorded in Municipality of Kuwait, 1965—67, Annual Reports, Section 8.
3. Base population as for Tables 21 and 22.

With such fundamentally different age-sex structures, the mortality experience of Kuwaitis and non-Kuwaitis also differs greatly. The very low crude death rates shown for the whole population 1958—1966 are a reflection of two factors—the under-recording of deaths and the extremely low general mortality rates of the non-Kuwaitis (Table 24).

aa) Kuwaitis

With such a young population very few Kuwaitis die of old age; in 1965 and 1966 almost one-third of the total deaths occurred amongst infants under one year old [19]. Text Figure 5 shows how the risk of dying is very high at birth, falling to low levels during the teens—a well-established world-wide pattern [20].

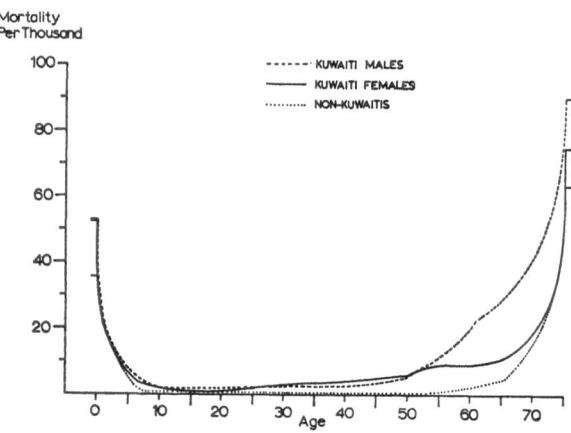

Text Fig. 5. Age specific mortality curves for Kuwaitis and non-Kuwaitis in 1965

That more Kuwaitis than ever before are surviving childhood is clear from the falling infant mortality rates. In 1965 the death rate of infants under one was 49.6 per thousand live births; by 1966 this had dropped to 39.5 per thousand. Infant mortality rates of under 25 per thousand are common in the developed world today. With continued heavy investment in health and education Kuwait can be expected to reach these lower mortality levels very soon.

bb) Non-Kuwaitis

The pattern of death varies amongst the non-Kuwaiti population from that described above. Broadly, early childhood and old age are still the most likely periods when death will occur, but age-specific death rates as a whole are much lower than for Kuwaitis.

As the summary table shows (Table 25), general mortality is low since there are proportionately fewer older non-Kuwaitis than older Kuwaitis living in Kuwait (Table 18). Deaths in these older age-groups occur in the migrants' countries of origin and are not recorded in Kuwait. However, infant mortality rates are also lower for non-Kuwaitis perhaps because of the higher educational standard and better health of the immigrants. This is difficult to substantiate statistically but in 1964 one-fifth of all infant deaths resulted from gastritis, enteritis, and colitis. Such maladies are often related to wrong feeding habits, themselves associated with illiteracy and lack of medical education [21—24].

4. Conclusions—Natural Increase

At this point, having reviewed the quality and the trends of the vital rates, we are in a position to reach several conclusions on the rate of natural increase in Kuwait. Table 25 summarizes the data presented so far.

Table 25. *A summary of vital rates for Kuwaitis and non-Kuwaitis in 1965*

	Kuwaitis	Non-Kuwaitis	Total
Marriages			
Crude marriage rate per thousand	—	—	4.3
Crude divorce rate per thousand	—	—	1.6
Births			
Crude birth rate per thousand	51.3	39.5	45.0
General fertility ratio per thousand	268	273	270
Deaths			
Crude death rate per thousand	9.8	4.0	7.3
Infant mortality rate per thousand	49.5	30.4	40.2
Increase			
Percent rate of natural increase	4.15	3.55	3.77
Females per thousand males	955	423	632

Calculated from:
1. For natality statistics, see Tables 21, 22.
2. For mortality statistics, see Table 24.
3. Marriage rates from Statistical Abstract 1967, Tables 17 and 18.

a) Kuwaitis

Superficially, Table 25 suggests that the Kuwaiti population is increasing as rapidly as any population in the world. Section 3 f has indicated several deficiencies in the mortality statistics, although the infant mortality figures are probably correct. If this is so, there seems no reason why the Kuwaiti population should not continue to increase at a high rate as infant mortality is further reduced to European levels (about 20 per thousand). The present rate of natural increase of 4.15 per cent per annum is probably a maximum figure because of under-registered mortality, but even assuming this high rate of growth, by 1970 Kuwaitis will number 269,700 and by 1975, 330,500.

What are the probabilities of Kuwaitis reaching these numbers? Several factors are involved, the most important being future government policy regarding the naturalization of non-Kuwaitis. If wholesale naturalization is accepted instead of the strict segregationist policy in operation at present (see above), obviously these numbers will be greatly exceeded.

Other factors include:
1. Fertility ratios have risen steeply recently (Table 21); these ratios may level off in future as female emancipation proceeds, or they may increase further as the per capita wealth also rises. Experience from elsewhere suggests the former outcome as more likely.
2. The 1965 Census population figures may be underestimates.
3. Contraception may become more widespread.
4. The Kuwaiti population will "age" as younger cohorts move into the upper age groups. Some reduction in the crude birth-rate can be expected as the proportion of women in the child-bearing age-groups declines for this reason (cf. experience in Hong Kong [25]).

b) Non-Kuwaitis

Despite the high natural rate of increase for non-Kuwaitis, the most important factor in their overall increase is the volume of immigration. The volume of immigration varies greatly from year to year—a net emigration of 17,800 in 1964 was followed by net increases of 12,800, 29,800, and 156,500 in 1965, 1966, and 1967, respectively (Statistical Abstract, 1967, Table 12; and Table 25 above). Assuming a net immigration of 20,000 per annum in future, almost an extra quarter of a million non-Kuwaitis could be in Kuwait in 1975.

By comparison, even with a natural increase rate of 3.77 per cent per annum, only an extra 103,222 will be added to the 1965 numbers of non-Kuwaitis by 1975. Recent trends observed above suggest that the crude birth rate of non-Kuwaitis may well surpass that of the Kuwaitis themselves. Numerically immigration will most likely remain of greatest significance in the increase of the alien population, but politically and economically the use of Kuwait's health and educational facilities by growing numbers of immigrant children may have a more profound significance.

VI. The Ecology of Daily Life

A. The Rural Community

Eastern Arabia offers very few opportunities for the development of a permanent system of agriculture. Evaporation and evapotranspiration far exceed the recorded precipitation which only occasionally rises above 150 mm annually. Along the eastern margins of the sand sea however, localized upwellings of fresh water in the province of Al-Hasa provide for the only sizeable concentrations of farmers between lower Iraq and the cultivators of Buraimi and the Sultanate of Muscat and Oman. In the oases of eastern Arabia—centres such as Qatif, Hofuf, or Al-Hasa itself—water is available in productive aquifers a few metres below the ground surface. The same water sources (derived ultimately from water percolating at deep levels eastwards from the better watered mountain areas of western Arabia) provide Bahrain with its onshore and offshore springs of fresh water, forming an important element in the Shaikhdom's growing economy.

Throughout Arabia, with the possible exception of the Yemen highlands, the search for water occupies a major part of the financial and technical resources of each country on the peninsula. In Saudi Arabia, for example, the search for water absorbed 74 per cent of the total agricultural budget in 1965—1966 [1]. In Kuwait the search for water continues with equal vigour for as yet, Kuwait's groundwater resources have proved inadequate even for the urban population without any allocation for agriculture. Brackish water from shallow wells coupled with the new freshwater field at Raudhatain constitute the total naturally occurring water resources of Kuwait, although survey parties are actively engaged in prospecting for further resources in the Wadi al-Batin in western Kuwait (Fig. 16). In short, the dense clusters of population in inland oases such as occur in eastern Saudi Arabia are not a feature of the rural areas of Kuwait. To a large extent this is due to the paucity of groundwater supplies, but a factor of major importance in Kuwait is the evolution of a cultural strain at variance with that of the settled agriculturalist. On the one hand are the urban-based commercial pursuits of the Kuwaitis and on the other the long-standing tradition of pastoral nomadism of the arid interior (Figs. 17 and 18).

The peripatetic mode of life associated with the grazing of herds of animals on a subsistence basis is closely linked to the territories on the margins of settled agriculture which are too dry, too steep, or too high to guarantee a worthwhile return to the cultivator. Special ecological difficulties are associated with the lands utilized for pastoral nomadism of which the foremost is moisture deficiency. Precipitation, particularly when amounts decrease in total, is an unreliable component of any agricultural system for the spatial distribution of rainfall, its total amount, and its seasonal incidence all vary considerably from year to year. On the arid and semi-arid margins these variations are magnified many times so that the pastoral nomad is in effect a gambler who pits his skills against the vagaries of the climatic elements. With only a limited technology at his disposal, the nomad's principal strategy to avoid drought for himself and his herds is to move—not in a haphazard fashion across unknown terrain, but proceeding along seasonal routes established by tradition as providing the least hazardous paths from summer to winter pastures and back again.

Beyond this it is very difficult to generalize, for each nomadic group (and there are many still surviving in the Middle East) has its distinctive annual routine and its concomitant systems of social organization [2]. One important factor which has special relevance in eastern Arabia is the relationship of the nomad to the farmers of the oases. Various systems of nomadism have been identified—viz. Merner's [3] division into "full nomads", "semi-nomads" and "mountain nomads"—and most of these classifications are too rigid to fit the diversity of ways of life which characterize the nomadic existence. Common to all types of nomadism, however, is the important reciprocal relationship which exists between the desert and the sown, for just as one relies on the agriculturalist for craft products and essential foodstuffs, so do the oases dwellers rely on the nomad for meat and meat products. In eastern Arabia, economic opportunities have traditionally been coastal in location (pearling, fishing, trading and piracy) which has meant that the nomads of Al-Hasa are closely linked both economically and through ties of kinship to townsmen or village dwellers. But even the traditions of the oil towns of the region are those of the desert—much altered, but still recognizeable as part of the strong cultural legacy of the Badu way of life. Contemporary Kuwait preserves many of these desert traditions despite the superficial veneer of westernism and for this reason justifies the examination of the nomadic existence, past and present, on the borders of Kuwait's settled areas.

The Badu of Kuwait. Documentary evidence on nomads and their seasonal routines is generally inadequate for a full understanding of the ecology of the Badu way of life. In Kuwait however, the interest of the one-time Political Agent Lt. Col. H. R. P. Dickson and his wife Violet in the Arabs of the desert surrounding Kuwait provides us with a valuable record covering all aspects of the nomad's pattern of life. This record is all the more valuable because of its compilation before the oil industry transformed the traditional movements of the Badu and spectacularly reduced the number of true pastoralists. The Dickson's books were published in 1949 [4], 1955 [5] and 1956 [6].

More than a dozen tribes made use of the good pasture land south-west of Kuwait, according to Dickson. Each had a tribal territory or *dirah* surrounding the orbit of their annual movements—which was shared with other groups with whom some alliance had been established. The majority of the tribes were camel herders but sheep and goats were also of importance. Dickson specifically cites five tribes—the Mutair, the Harb, the Shammar, the Awazim, and the northern Ajman—who traditionally made use of Kuwait town as their summer base. From June to early October the tribes camped around the shallow wells just outside the walls of Kuwait town enduring the intense heat in their black tents made from goat hair. This period of inactivity was a frustrating interlude for the Badu who were vulnerable to raiding parties from rival groups and to the Ruler's tax gatherers seeking payment of the annual tribute or *zakat*. Early air photographs of Kuwait in summer show the black tents clustered around the shallow wells at Shamiya and also inside the walls near the present district of Mirqab (Fig. 19). Rows of untidy huts can be seen just to the south of Safat Square where the *Badu* gathered daily to exchange news in the course of their shopping expeditions in Kuwait's market or *suq*.

With temperatures dropping gradually in October, the rising of the star Canopus was the sign for the tribes to strike camp and to move off into the desert (Fig. 20). Initially grazing was very scarce and the first heavy showers of rain were greeted with great rejoicing. Moving tents every ten days or so for sanitary reasons and because of the exhaustion of the grazing, the tribes moved up to 200 miles into the desert across the Dhana sands into eastern Saudi Arabia. The direction of movement was based on news obtained from neighbouring tribes and on careful assessment of the state of the grazing by the head of the group. One of the largest tribes—the Mutair—described a roughly oval path due south from Kuwait and then swinging westwards across the Dahna to the wells at Anaiza. By May, the mounting temperatures signalled the moment when once again the tribes should converge on the towns of eastern Arabia to sit out the long uncomfortable summer.

Detailed descriptions of the tribal organization, the daily routine, and then Badu philosophy together with sketches of their tents and standard household equipment are provided by Dickson [7] and need little elaboration here. Some of the most important aspects of the Badu tradition need some emphasis because of their carry-through into the social organization of modern Kuwait.

Above all else, the people of Kuwait are Muslims mostly belonging to the Sunni or Orthodox schism of Islam. Wahhabism, a puritanical movement particularly associated with the rise to power of the House of Saud early this century, still makes its influence felt in present day Kuwait. Basing their philosophy on the literal teaching of the Qu'ran, the Muslims of Kuwait—many of them Badu converted to Wahhabism during Ibu Saud's recruiting drives amongst the nomads of eastern Arabia—still insist on a total ban on the open sale of alcohol in Kuwait, and on the closure of restaurants and cafes during the daylight hours of Ramadhan. Similarly, several Kuwaitis still shun usury, including under this heading receiving interest on their investments, much to the dismay of the bankers of the State. To this day, veiled women are a common sight in the market of Kuwait, the black 'abbas or cloaks concealing comely European-style dresses and shoes.

Stemming from the religious traditions characteristic of the desert rather than the towns of eastern Arabia, there follows a whole selection of practices and customs which are maintained in the modern city of Kuwait. Many of these traditions centre around the Holy Month of Ramadhan when smoking, eating, or drinking in public is strongly discouraged. At night however, when the gun is fired to signal the setting of the sun, the streets are bare as groups of friends and relations meet to celebrate the "Iftar" ("breakfast") with an enormous meal. After the meal, the arduous round of visiting begins first to the immediate kin and then beyond to friends and more distant relations. Cafes and shops stay open well into the night, attracting crowds of cheerful people parading up and down the brightly lit arcades and pavements. At the end of the month of fasting, the day of the 'Id al-Fitr—a feast and a general holiday—is announced based on intelligence received from Mecca about the sighting of the new moon. Once again the round of visiting begins, starting with the spontaneous flood of people who call on the Ruler and wish him a "Blessed 'Id" ('Idkum mubarrak). Soon the season for the Haj or pilgrimage approaches. Instead of the tiring trek on camelback across the deserts, the pilgrims can now fly to the Holy Cities or join a caravan of packed buses on the all-weather road laid from Kuwait right across the peninsula of Arabia. Leave is automatically granted to Civil Servants in Kuwait wanting to make the journey and Kuwait's contribution of funds and medical missions which are situated along the pilgrimage route ensures a safer and more comfortable Haj for all concerned.

The second feast, the 'Id al-Adha, marking the culmination of the pilgrimage in Mecca, is celebrated with equal energy in Kuwait. For the children, it means a flood of presents and gifts; for the parents it usually means a short holiday—perhaps a trip abroad. Kuwait's borders are crowded with outgoing holiday-makers at the 'Id—both the airports and the land frontiers experiencing a surge of travellers. At home, tents are set up on open spaces in the city and groups of men gather to sing and dance in the traditional manner to the accompaniment of negroes beating drums and banging small cymbals. Clearly the major elements of the religious calendar are unlikely to decline in overall significance if the present gusto of the celebrations is maintained.

Apart from the religious side of life in Kuwait, several other prevalent customs testify to the importance of the contribution which the desert makes to the everyday routine of the modern city. A prominent feature of tribal life is the emphasis on kinship; in Kuwait, families still maintain their connections (reinforced during the 'Id's and in Ramadhan) to such an extent that several major groups of families hold positions of enormous wealth and power in the state. Examples of such families are the Al-Khalids, the Al-Ghanims, the Al-Saqars, or the Jana'at grouping—names which crop up amongst Cabinet ministers and leading businessmen. At a more lowly level, families maintain their links—even when a trip of several kilometres is involved in paying a routine call. Even to callers outside the immediate family circle, the traditional hospitality of the Badu is freely extended. Calling on officials in government offices involves sharing a glass of tea, a cup of coffee, or even a less traditional carbonated cold drink. The same stylised greetings are used by Kuwaitis in their air-conditioned offices as the desert dwellers extend to their visitors in their goat-hair tents. This follow-through of past modes of life in a novel economic context is a characteristic of Kuwait and indeed all the small Gulf states of eastern Arabia. To a large extent the anachronistic traits of the desert account for much of the foreign visitor's fascination with Kuwait. But today Kuwait is an almost totally urbanized nation of an astonishing complexity: many of the traditions of social organization stem from the desert but much of the urban fabric and its various functions are entirely novel.

B. The Urban Community

Post-War Expansion of the Built-up Area

Following the rapid rise in oil revenue in the post-war period and the sizeable influx of foreign-born immigrants into Kuwait (Chapter V), increasing demands were placed on the Old City which it was ill-equipped to withstand. To relieve both traffic congestion and residential squalor (see below) the Government embarked on a rapid programme of urban development in the 1950's which largely shaped the form of the present-day city.

1. Kuwait City Before Oil Discoveries

a) Extent

Early accounts of the form and extent of Kuwait City are sparse; one of the very few 19th century travellers to visit Kuwait was J. H. Stocqueler in 1831. He writes of the city:

"Koete, or Grane as it is called in the maps, is in extent about a mile long, and a quarter of a mile broad. It consists of houses built of mud and stone, occasionally faced with coarse *chunam*, and may contain about 4,000 inhabitants. The houses, being for the most part square in form, with a court-yard in the centre (having the windows looking into the yard), present but a very bare and uniform exterior, like, indeed, all the houses in the Persian Gulph. They have flat roofs, composed of the trunk of the date tree. The streets of Koete are wider than those of Muscat or Bushire, with a gutter running down the centre. A wall surrounds the town on the desert face, but it is more for show than protection, as it is not a foot thick" [8].

Lorimer's "Gazetteer" of 1908 provides more substantial information on the form and extent of Kuwait City—at that time the only centre of significance is Kuwait. The city in the opening years of this century extended 2 miles along the shore and a quarter to three-quarters of a mile inland in a rough semicircle. Between 1870 and 1900 the population of the city and the built-

b) Internal characteristics and differentiation

Arab cities are noted for their familial affinity in structure and architectural form. Kuwait, before reconstruction began in recent years, had several attributes shared by Middle Eastern cities elsewhere. Lorimer mentions names of quarters in Kuwait and refers to the dif-

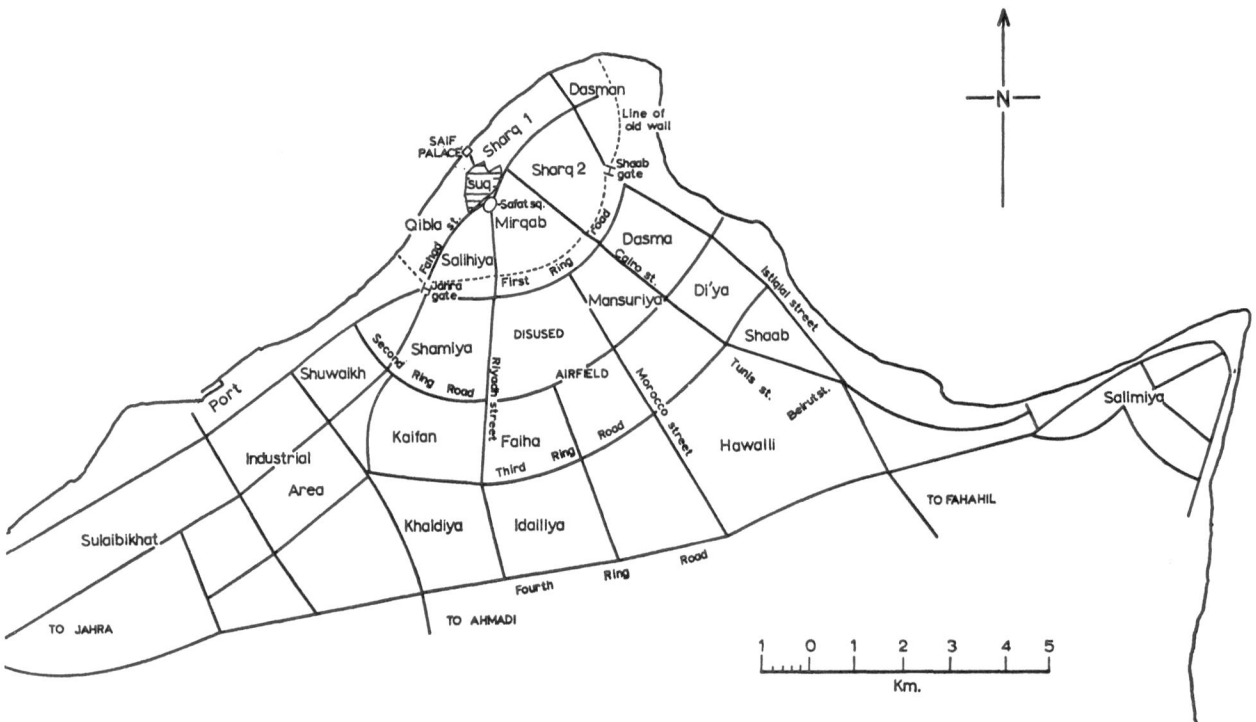

Text Fig. 6. Census divisions and place names in Kuwait City

up area had doubled, probably by immigration. Sites granted by the Shaikh were rapidly being built upon, particularly in a long suburb on the south-western edge of the city called "Mirqab". City development was continuous and close-packed:

"The streets are irregular and winding, and many of them blind alleys and the town is not laid out in any general plan; the only street of apparent importance, besides the main bazaar which runs at right angles to the sea about the middle of the town, is one which leads from the suq or market square situated at the back of the town near the Mirqab quarter, to the north-east of the town but it has no general name. Most of the houses have only a ground floor but appear higher owing to the parapet wall enclosing the roof; they are generally built surrounding a courtyard" [9].

Subsequent descriptions are rare up to the modern period but photographs become more common, especially when oil exploration began in earnest in 1935. Vertical air photographs are not available before 1951, but both K.O.C. and the Government have a selection of horizontal and oblique photographs dating from the 1930s. Figs. 21 and 22 illustrate the high density single-storey housing pattern which persisted until the 1950s in most of the town and which remains today in some older parts of the city (e. g. Sharq on the sea front). Fig. 23 shows the broadly semi-circular outline of the city, particularly the space which separated the continuously built-up area from the outer wall constructed in 1920. This space was gradually in-filled by later settlement, particularly by the southern extension of Mirqab quarter (Text Fig. 6 and Fig. 27).

ferentiation between the residential areas for the 4,000 Negroes (two-thirds of whom were still enslaved or *Mamluk* in 1907) and those for the other Arabs. Without knowing more about traditional society at that time, it is difficult to enlarge further on functional areas within the city, but several distinct structural areas are discernible on the early air photographs.

1. As Shiber [10] points out, the central square (*Maidan* or *Saha*) is the focus of most Arab cities. The central square of Kuwait called Safat Square since mid-1930s has always been a notable feature of Kuwait City. The square was the departure and arrival point for caravans crossing the desert since the Custom House was located near the square, together with the cafes consisting of tin and rush huts. These cafes, according to older Kuwaitis who remember this period, were the meeting places of Badu and merchants who gathered in Safat Square to conduct business and renew acquaintances.

2. The commercial core of most Arab cities was the *suq* or bazaar, preserved in many cities such as Kuwait where it still concentrates a sizeable proportion of the retail and wholesale trade. Figs. 25 and 26 show the original bazaar stretching from Safat Square towards the dhow harbour and the Ruler's Palace on the waterfront. This block between Safat and the sea was a maze of narrow twisting lanes, some of them roofed with rush matting or corrugated iron, each lane containing a row of small stalls, mostly selling closely allied goods. In Fig. 26 the corrugated iron roofing, erected during World War II, indicates the layout of the main shopping lanes.

Near Safat, the small shops still specialize in Badu requisites—weapons, leather goods, tent material, cloaks, and rope. Further towards the coast was the next market. Much of the *suq* as a whole remains today; districts within the *suq* are still known by the name of the principal products sold, e. g. *Suq az-Zil*—the carpet suq; *Suq as-Silah*—the weapon suq; and *Suq al-Laham*—the meat suq.

3. Mosques are prominent features of settlements throughout the Islamic world. Early this century Kuwait had between twenty and thirty, four of which were Friday *jami* mosques. In Figs. 27 and 28 the squat minarets can be seen scattered throughout the city, each mosque serving the needs of its nearby residents. Many of the mosques were family mosques constructed by a group of relatives in a particular quarter. The largest mosque, attended by the Ruler on public feasts and festivals, is *Masjid as-Suq* located in the centre of the bazaar (Fig. 23 centre).

4. Several of the Ruling family and the richer merchants constructed "palaces"—in fact, large houses—in the years before oil exploration began. The Ruler himself maintained a walled residential enclosure at Dasman (Text Fig. 6) and a Council Chamber in the town centre on the waterfront—now the Saif Palace (Fig. 29). Along the shore between Dasman and the Saif Palace richer members of Kuwait's society built substantial two-storey houses which today are still prominent features of the Kuwait waterfront.

5. Finally, as is common with most Arab countries, graveyards in Kuwait are regarded as inalienable plots of land. On Fig. 27 on the 1951 air photograph, the large open spaces within the built-up area are graveyards—several of them disused—on which residential and other development is forbidden. In only one instance—the graveyard north-west of Safat Square—has development been permitted, in this case the construction of a public park (Fig. 28). Recent legislation will allow the development of these older cemeteries in future, thus erasing a prominent feature from Kuwait's townscape.

2. Kuwait City after the Discovery of Oil

With spiralling oil revenues and a rising immigrant population Kuwait City expanded rapidly both vertically and horizontally. Expansion was not haphazard, as the degree of financial and legislative power which the Government was able to exercise was impressive by any standards (see Shiber [11], for details). These powers were freely used to transform the city physically, incidentally beginning a process of population re-adjustment in which people were moved from within the Old City to the newer suburbs beyond. Air photographs taken in 1951, 1960, 1964, and 1967 are invaluable evidence in tracing the course of this post-War phase of planned urban expansion.

Fig. 27 shows the entire built-up area of Kuwait City in 1951 forming a rough oval centred around the Ruler's Palace and the docks on the waterfront. Clearly visible are the features mentioned above—the central square called Safat, the covered suq running between the sea and Safat, the tiny scattered mosques and minarets, the larger houses on the waterfront, and the open spaces of the graveyards—all enclosed by the semi-circular wall built in 1920.

By 1951, despite five years of oil exporting and revenue payments, little physical change had overtaken Kuwait. Immigration was beginning on a large scale but accommodation was provided by K.O.C. for Company employees at Maqwa or Ahmadi, while other arrivals lived in shacks and tents on the periphery of the city (Fig. 27) or west of the city in a labourers' camp (Madina al-Amal). However, both the growing number of motor vehicles (over 1,000 in 1951) and the number of new arrivals were good reasons for embarking on a programme of wholesale city development. There was another reason—dealt with in detail below—which was to distribute the oil wealth throughout the private and the public sectors of the economy. Large-scale purchase of land in the Old City by the Government, together with the benefits brought to the merchant community through contracts for Government constructions projects, were important facets of the fiscal policy of spreading the oil wealth amongst Kuwaiti citizens. This land and property acquisition scheme warrants close attention not only because of its effects on the economy as a whole, but also because of the changes which it wrought in the structure of the city. Land values are fundamental components of the structure of any city, so that a detailed study of the property acquisition scheme was undertaken with the aim of identifying the structure and changes in Kuwait's property values from the inception of the scheme to the present day.

3. The Government Land Purchase Scheme

The property acquisition scheme began in 1951 with the twin aims of infusing sums of money into the economy's private sector whilst at the same time facilitating the wholesale reconstruction of the Old City. Owners of land and property within the City were offered deliberately inflated prices by the Government to encourage the owners to move out into the newer suburbs and at the same time to provide these Kuwaitis with a certain amount of working capital. Table 26 indicates the amounts of money involved; almost K.D. 600 million was disbursed in less than 15 years.

The Mission of the International Bank for Reconstruction & Development expressed some criticism of the scheme. In the six years before 1964, 50 per cent more had been spent on property acquisition than on public capital projects [12]. Much of the money disbursed on the scheme is apparently remitted abroad in private in-

Table 26. *Annual expenditure on the government property acquisition scheme in Kuwaiti Dinars between 1952 and 1967*

Year	K.D.	Year	K.D.
1952	2,142,088	1960—61	42,964,939
1953	3,431,052	1961—62	58,858,578
1954	6,836,709	1962—63	46,472,873
1955	12,684,179	1963—64	31,537,287
1956	12,139,907	1964—65	44,997,006
1957	21,628,181	1965—66	77,801,679
1958	40,147,909	1966—67	95,000,000[b]
1959—60	85,335,176[a]	Total	581,977,563

[a] Financial year of 15 months.
[b] Estimate for the calendar years of 1967: up to March 13th 1967, K.D. 72,650,174 had been spent.
Source: Ministry of Finance, March 18th 1968 (Gazette).

vestment portfolios and is thus of relatively little utility in the economy of the State as a whole.

Nevertheless, the overall success of the scheme is inescapable, both in hastening Kuwait's economic development and in permitting the very rapid re-development of the Old City. One of the most important effects of the scheme was the way it brought about the amalgamation of tiny plots of land in private ownership into larger state-owned blocks suitable for re-development.

Table 27. *Average annual prices in Kuwaiti Dinars per square metre for property in the old city*

Year	Land with house	Land only	Year	Land with house	Land only
1952	—	4.033	1960	162.370	129.203
1953	12.917	7.937	1961	159.254	—
1954	28.062	14.133	1962	145.099	—
1955	48.581	10.899	1963	207.228	—
1956	48.121	23.309	1964	175.070	—
1957	77.696	19.375	1965	187.360	—
1958	71.256	37.944	1966	199.359	—
1959	221.370	68.622			

Calculated from: Municipal Property Records.

Prices for land rose 32 times over between 1952 and 1960, while those for houses increased 15.4 times over between 1953 and 1966. Neither showed a steady increase because of the "skewing" effect produced by the purchase of large blocks of either very dear or very cheap property in any one year. In addition, the Valuation Committee considered that the rise in prices in 1960 was too steep (between 1958 and 1959 prices tripled) so that prices for houses and land were lowered by 25 per cent and prices for open ground by 15 per cent. By 1963, however, prices had surpassed their 1960 level.

4. Planning and Urban Expansion

a) The 1952 plan

Clearly Government planning was bound to play an important role in shaping the form of the new city with the Administration assuming such an important economic position through the property acquisition scheme. In 1952 a British Consultant firm, Minoprio, Spencely & Macfarlane, was enlisted to produce a master plan which was to determine to a large extent the location of subsequent development and the form of the contemporary city. This Plan involved the Government in the construction of 17 major residential blocks (A to Q on Text Fig. 7), arranged concentrically around the Old City and linked by a broadly radial road system. Hawalli was to be expanded to become a town of 15,000 people outside the grid-iron inner neighbourhood blocks. Shuwaikh (areas R and S) was reserved for industry, and areas T, U, V, and W west of Shuwaikh for health and recreational land uses. Despite suggestions for the construction of an entirely new city outside the old town (Ministry of Guidance, 1963, p. 148; Shiber, personal communication), the Plan recommended that the town within the wall should remain the city centre.

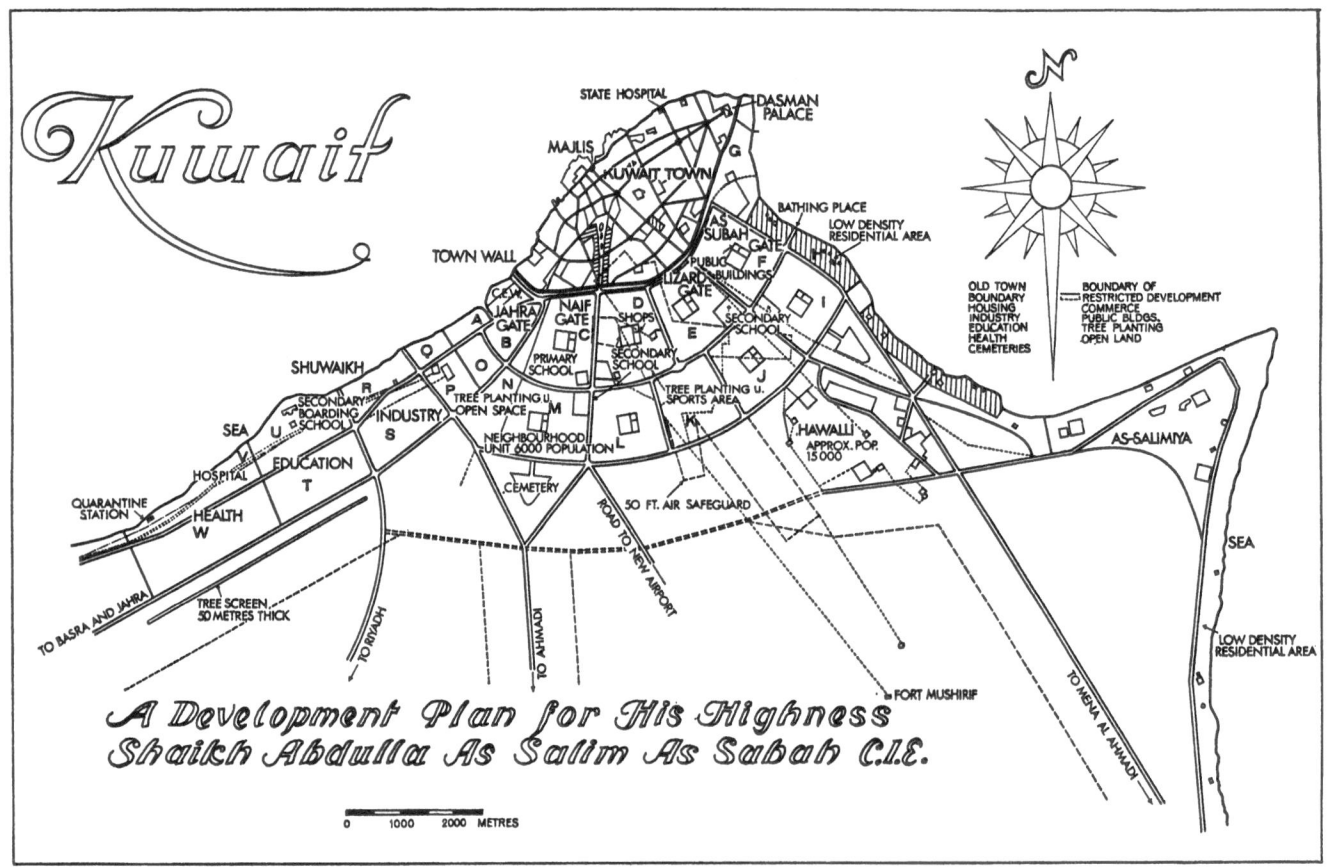

Text Fig. 7. The 1952 development plan for Kuwait City. After Kuwaits' Public Works Department Drawing No 387/36

Thus the Plan embodied several points of lasting significance to the final form of Kuwait City:

1. Segregation of Kuwaitis and non-Kuwaitis was established (by the Government and not by the Plan), with the creation of major new neighbourhood blocks (A to Q on Text Fig. 7) to which only Kuwaitis were transferred as the reconstruction of the Old City proceeded during the 1950s. Less formal plans were laid for the expansion of Hawalli and Salimiya, later to become important areas of residence for non-Kuwaitis.

2. The Old City was scheduled for wholesale redevelopment as a modern city centre serving all of "Greater Kuwait". Only secondary consideration was given to residential accommodation within the wall (Fig. 31).

3. Throughout the built-up area the road system was to comprise a series of radial dual-carriageways with semi-circular cross connections, together with a system of minor roads in a broadly rectangular layout within the new neighbourhoods. Overall, the traffic system was to be highly centralized on the Old City.

4. Each new neighbourhood was to have a selection of centrally located services—schools, shops, mosques, etc.—which were designed to meet the local needs of that district's residents.

b) Physical expansion of the city structure

Once the guidelines of future urban growth had been laid, construction of the new suburbs proceeded rapidly. Text Fig. 8 indicates the extent of the built-up area be-

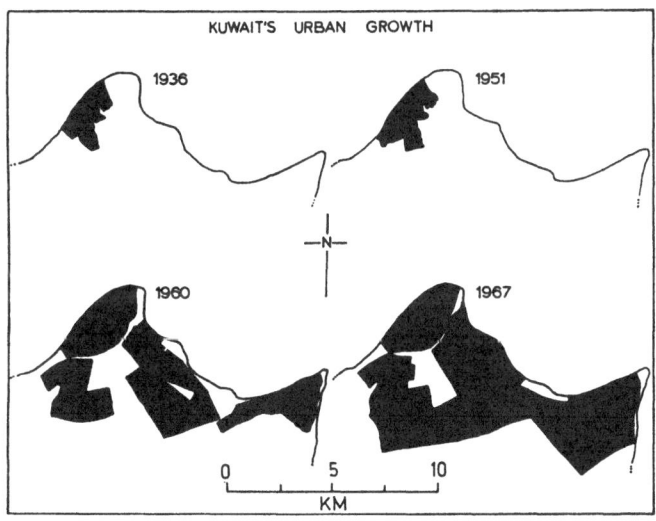

Text Fig. 8. Expansion of the built-up area of Kuwait City between 1936 and 1967

tween 1936 and 1967. Initially suburban development took precedence, but in the late 1950s attention was directed towards the transformation of the Old City, particularly the erection of large public buildings. Shiber [13] describes this period of frenetic building activity in detail; construction followed planning with unprecedented haste. Fig. 27 and 28 show the Old City in 1960, and 1967, indicating a picture of apparent havoc and desolation. Most of the Old City in its original form is to disappear, but one recent report [14] strongly criticises

this replacement of the Old City with mediocre new buildings.

Within the Old City building took place in the open spaces bordering the built-up areas of the 1951 City before urban renewal began on a large scale within the City itself. Parts of Mirqab were quickly appropriated for the Ministry of Public Works, while commercial development in Salihiya began in the late 1950s and early 1960s. Only in the mid-1960s were extensive inroads made into the rebuilding of the centre of the Old City where new commercial centres are at present emerging on the seaward side of the Suq, opposite the Saif Palace. A substantial concentration of public buildings—the Ministries of Social Affairs & Education—is emerging in the Commercial Area 9 immediately to the east of Safat Square.

With some of the new suburbs up to 10 km from the city centre, new problems of movement have arisen within Kuwait's built-up area. In 1966, 80,331 private cars and a further 14,577 public vehicles were registered in the State, amounting to about one vehicle per household. As a result, traffic problems have reached significant proportions in parts of the Old City and at major intersections.

The advice of another firm of consultants has been sought on these and related problems. Future urban planning is likely to be closely co-ordinated with plans for the expansion of the economy as a whole, and since industrial diversification is a major goal of this expansion, the development of a heavy industrial area in south Kuwait is of some importance in the location of new residential areas. Until recently Ahmadi was the only town of over 10,000 inhabitants in south Kuwait. Figs. 31 and 32 illustrate the rapid genesis of Ahmadi from a mere oil company camp to a fully fledged town at present approaching 20,000 in population. Fahahil has experienced a swift population increase from 8,923 in 1957 to 20,782 in 1965. Shuaiba too, designated a future industrial area with its commercial port, oil refinery, and petro-chemical plants, is also expanding rapidly both in population and the extent of its built-up area. As yet, the new town of Sabahiya between Ahmadi and Fahahil has not emerged as a physical reality. Thysse's recommendations have an important bearing on the balance between developments in south Kuwait and in the capital.

c) Subsequent plans

Whilst Kuwait's urban expansion was almost unfettered throughout the 1950s, by the early 1960s almost all of the 1952 Plan's aims had been realized. With the city closely approximating to the outline shown in Text Fig. 7 outside Consultants were again asked to advise on future development. Thysse, reporting in 1962, wrote:

"My expectations of a large population in Kuwait City are not very high. The population number will not be very much increased and consequently in the near future Kuwait City will not grow very much larger than its present size." [15]

Instead, accent was laid on the industrial and residential development of south Kuwait, centred on the Ahmadi—Fahahil—Shuaiba complex. This aspect of Thysse's recommendations is nearer fulfilment than his recommendations concerning Kuwait City for two main reasons:

1. Thysse clearly overestimated Kuwait's potential for industrial growth. While he attributed most of Kuwait

City's "propulsion" to the building industry [16], as Chapter 4 has shown, Kuwait's major employment category is the provision of services. Ahmadi's future industrial development is unlikely to alter this service bias of the economy for several years.

2. The volume of immigration to Kuwait is almost unpredictable. Thysse could not have foreseen the political upheaval of June 1967 and its effect on Kuwait's total population growth. As indicated below, the immigrants show a marked preference for Kuwait City.

The most recent town planning consultants' report concentrated on the need for a comprehensive Master Plan, emphasizing the poor architectural standards in Kuwait and the seriousness of traffic problems, especially in the Old City [17]. At present other consultants are working on the formation of an overall Master Plan, but its effect on the form and functioning of the present-day city are as yet not apparent.

5. The Contemporary City

Overall, urban centres in Kuwait today represent the outcome of nearly 20 years of controlled development. Kuwait, more than most states, has an almost completely planned environment. What does this environment comprise?

a) The Old City

Most of the Old City has been purchased by the Government, razed and re-developed. Parts of the 1951 city remain (Fig. 28), including sections of Sharq, Mirqab, and the suq. Elsewhere, most of the street facades comprise modern concrete building even though the areas behind are as yet undeveloped. Notable additions include Fahad as-Salim Street (Fig. 33)—an entirely new shopping district specializing in "modern" shops; several commercial centres near the suq; a banking district on the seaward side of the suq; and a Government Administration concentration based on the Ministry of Public Works Compound in the south-east section of the city near the Shaab gate. Throughout the Old City, wide dual-carriageways have replaced the old lanes, and on the seaward side a corniche road in course of further enlargement has radically altered the northern face of the city.

b) The Kuwait neighbourhoods

Much of these suburbs comprise roughly rectangular blocks 2 sq. km in area built around a central square containing the neighbourhood service centre (supermarket, mosque, medical clinic, social centre, schools, and a parade of shops). Houses are built on 750 or 1,000 sq. m lots from concrete, and are surrounded by a brick wall over 2 m high for privacy. Despite the variety of architectural styles, an austere impression is created by the uniformity of construction materials, the formality of the road pattern, and the high surrounding walls reflecting the neighbourhoods' evolution as a result of closely controlled planning.

c) Hawalli and Salimiya

By comparison, Hawalli and Salimiya have evolved more freely since these districts are permitted places of residence for non-Kuwaitis. Almost one-third (74,500)

of the immigrants lived in these two districts in 1965, outnumbering the Kuwaitis by almost 3 : 1. Hawalli and Salimiya have a greater variety of layout and building styles than the other neighbourhoods because of speculative building by Kuwaitis on behalf of the immigrants.

d) Abruq Khaitan and Farwaniya

Situated beyond the continuously built-up area, these two districts have developed rapidly as dormitory towns for non-Kuwaitis. 26,000 non-Kuwaitis lived there in 1965 compared with 18,000 Kuwaitis (Census of Population, 1965; Table 1). Although formally laid out like the Kuwaiti neighbourhoods, Abruq Khaitan and Farwaniya, like Hawalli and Salimiya, have developed more freely as both residential and retail developments are less strictly controlled.

Jalib as-Shuyukh beyond Abruq Khaitan is largely a Badu settlement comprising older plaster buildings and hundreds of tin and rush matting huts. "Low income" houses are in course of erection.

e) Ahmadi and Fahahil

Ahmadi is a completely planned centre, but it differs in appearance from Kuwait City's neighbourhoods because of its bungalow-style of housing and the stress placed on tree and shrub planting by K.O.C. (Figs. 34—37). Fahahil is a marked contrast because of its untidy sprawl outwards from the core of the older fishing village. Residential developments at Shuaiba are as yet limited to huts and a few older houses.

C. Population Distribution and Density within the Urban Areas

Kuwait's population has never been widely scattered over a rural area, instead, Kuwait City has always contained at least two-thirds of the State's total population. Indeed, at the time of Lorimer's visit, 35,000 people lived in the City compared with a total population of 37,000. Little change in this highly concentrated pattern of distribution was registered up to the 1950s when oil and associated developments in south Kuwait began to produce sizeable population agglomerations beyond the built-up area of Kuwait City. Fortunately much of the important period in the later 1950s and throughout the 1960s, when Kuwait City was being expanded and rebuilt, and when the Ahmadi-Fahahil complex was emerging, is covered by the three Censuses of Population in 1957, 1961, and 1965. These statistics provide us with the information required to consider changes in the overall population distributions; the hierarchy of centres in Kuwait; and density gradients within the urban areas.

1. Changing Patterns of Population Distribution

a) Before the first census

At the beginning of the century the area tributary to the Shaikh of Kuwait's suzerainty (roughly equivalent to the area of the modern State, excluding some date gardens at Fao) contained 37,000 people with an additional 13,000 migratory Badu. Of the 37,000, 35,000 lived in Kuwait City. Jahra was the only other centre of any size, containing 500 people—swollen in summer to 700 by the incursion of nomads. Eight smaller settlements were recognized, all of them smaller than Jahra— Qasr as—Sabiyah; Failaka; Dimnah (called Salimiya from 1953 onwards); Sirrah; Fahahil; Fantas; Abu Halifa; and Shuaiba [18].

This pattern of population distribution—highly centralized in Kuwait City with very minor population concentrations on the east coast and at Jahra—was preserved until the post-1946 period. Kuwait City grew steadily between 1908 and 1946; because of the exceptional factors affecting this growth, it is most unlikely that any other centre apart from Kuwait City experienced any notable increase in the first half of the twentieth century.

b) Effects of oil discoveries

Up to 1946 K.O.C.'s labour force did not exceed 300, but in the post-war period labour needs rose sharply from 1,552 in 1946 to almost 8,000 in 1950. As indicated above, most of the labour force and its dependants resided and worked in south Kuwait, first at Maqwa and then finally at Ahmadi, the present-day base of all K.O.C.'s activities. Initially, the numbers of dependants were small since the census population of Ahmadi in 1957 was only 7,280 compared with K.O.C.'s total labour force of 9,038 for the same year. Fahahil, with a population of 8,923 in 1957, undoubtedly benefitted from the discovery of oil in south Kuwait, but together with Ahmadi the growth of these centres constituted the first major shift in the population distribution since the predominance of Kuwait City had been established almost two centuries earlier.

c) Population distribution in 1957

Text Fig. 9 illustrates the results of the first population Census. As well as the decentralization effect which the oil industry had on population distribution, another powerful force which caused a marked re-adjustment of population was the reconstruction in the Old City. As shown above, new suburbs were quickly laid out and constructed according to the design recommended by the first Master Plan, while at the same time work began apace on the wholesale reconstruction of the Old City within the wall.

While the distribution of the total population in 1957 conforms to an apparently straightforward pattern (Text Fig. 9), population movements within Kuwait are complex. Two factors are of prime significance:

1. Most of the new suburbs were reserved for Kuwaitis only. Hence the evacuation of the Old City was a movement involving Kuwaitis only, while most of the new immigrants continued to flock into the city centre.

2. Rapid increases in the volume of immigration throughout the 1950s coupled with Kuwait's strict nationality laws produced a dualism in population distribution preserved until the present day. Much of the homogeneity of the population was lost, since aliens and citizens possessed contrasting demographic characteristics and different places of residence.

A measure of the magnitude of these factors is provided by Table 47 of the 1957 Census which lists

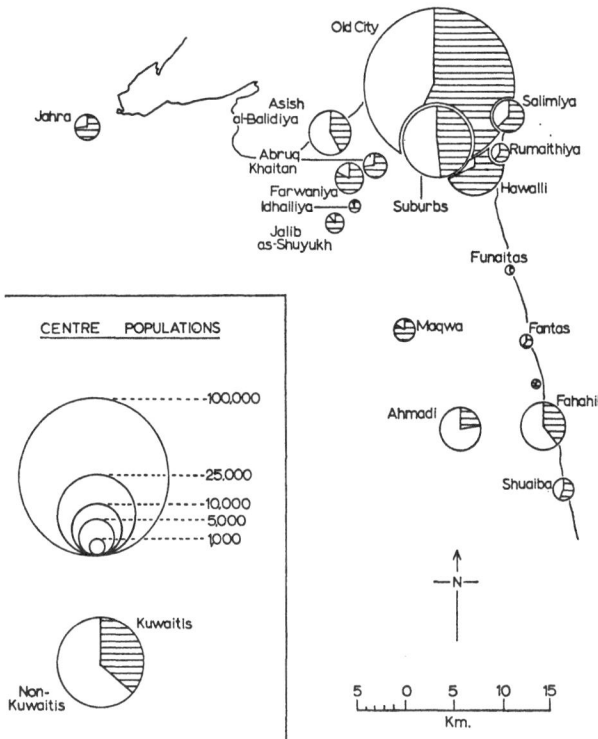

Text Fig. 9. Population distribution in 1957. At the outset, Kuwait City was by far the largest Centre in Kuwait

Kuwaitis and non-Kuwaitis separately by place of birth and place of enumeration. Text Figs. 10 and 11 have been constructed from this table. Clearly shown is the movement outwards into the suburbs of Kuwaitis born inside the Old City; altogether, 32,962 Kuwaitis were involved in this movement while 55,126 remained within the Old City—their place of birth (Census of Population, 1957, Tables 47a and b). Only 3,327 Kuwaitis born in the Old City moved to Ahmadi or Fahahil before 1957; evidently the capital in its new form provided too many attractions (both housing facilities and employment opportunities) to cause any significant volume of outmigration to south Kuwait.

By comparison, the non-Kuwaitis showed a much more diverse pattern of movement within Kuwait. Of the total of 83,348 non-Kuwaitis shown in the census, 38,888 resided in the Old City. Only 2,153 of these residents had been born in the Old City, so that this district of Kuwait was the most popular single target with incoming migrants. As Text Fig. 11 shows, other popular centres for migrants born outside Kuwait were the new suburbs (10,298 non-Kuwaitis lived there in 1957); Fahahil and Ahmadi (9,615); Hawalli (4,302); Ashish al-Baladiya (3,756); and finally Salimiya (1,421). The high proportion of non-Kuwaitis residing in the suburbs (10,298) is misleading; most of these people were male Iranis and Iraqis (4,488) employed as labourers in the construction of the new suburbs and living in huts on-site. Ashish al-Baladiya comprised a shanty town hastily erected to hold the large numbers of migrant workers flocking into Kuwait during the early 1950s.

Text Fig. 9 summarizes the distribution of population in 1957 resulting from these movements and expresses the proportion of each centre which was composed of non-Kuwaitis (immigrants).

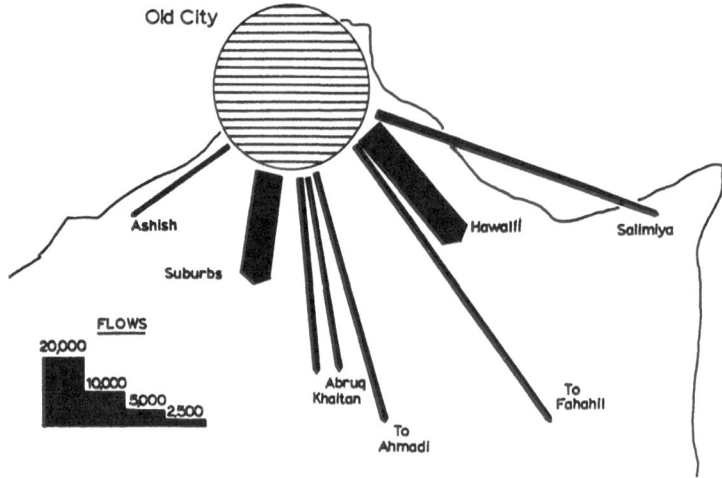

Text Fig. 10. Kuwaitis by place of birth and place of enumeration in 1957

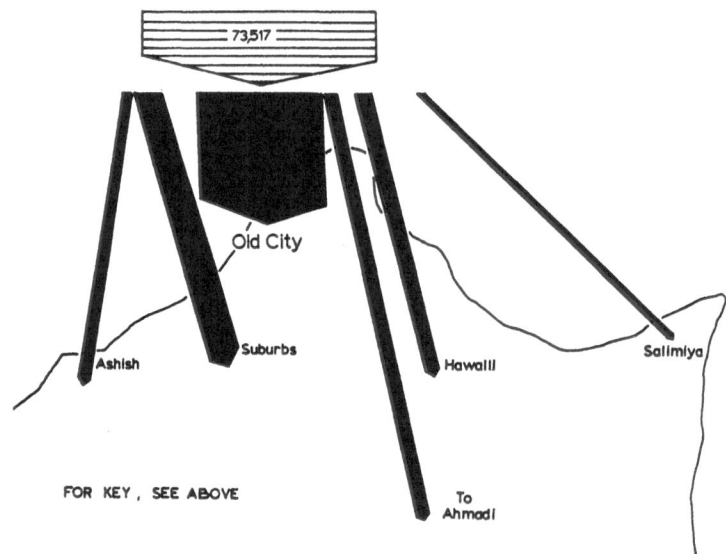

Text Fig. 11. Non-Kuwaitis by place of birth and place of enumeration in 1957

d) Population distribution in 1965

As Text Fig. 12 shows, by 1965 Kuwait City entirely dominated the population distribution of Kuwait with 298,701 out of the total of 467,339 people living in the area of "Greater Kuwait" (the continuously built-up area within the area circumscribed by the fourth ring road—Text Fig. 6). Particularly interesting is the rapidity of the expansion of the areas not specifically reserved for Kuwaitis. Immigration has added proportionately more to the total population than natural increase and as a result residential areas open to non-Kuwaitis have sustained a high rate of growth in the inter-censal period 1957—1965. Particularly noticeable in this context is the growth of Salimiya (4,075 to 38,648 from 1957 to 1965); Hawalli (14,784 to 54,542); Fahahil (8,923 to 20,782); Abruq Khaitan (2,153 to 23,610); and Farwaniya (3,261 to 20,444). These latter two centres have acted as "overspill" towns for Kuwait City because of their location just 4 km south of the fourth ring road.

With heavy immigration continuing in the post-1956 period these centres are likely to have experienced further large population increases while the populations of the exclusively Kuwaiti neighbourhoods are more likely to remain almost constant. No statistics are available on the place of residence of non-Kuwaiti immigrants arriving after 1965, but further overall increases will probably exaggerate the dichotomy in the physical and social attributes of the Kuwaiti and non-Kuwaiti districts.

D. Population Composition and Social Areas within Kuwait

Up to this juncture city structure has been referred to but not discussed at length. The physical expansion of Kuwait City was examined above together with some aspects of the structure of the contemporary city; however, there is an entirely different aspect of city structure which includes the social, demographic, and economic characteristics of districts within the city. Most of the established theories in urban geography relate to the distribution and change in these and similar variables, themselves closely associated with the physical attributes of an area.

1. "Western" and "non-Western" Cities

Frequently in this work allusions have been made to the differences between cities in the countries of North America and Western Europe and those elsewhere in the world. Several terms have been used to differentiate the two groups—Berry [19] uses "Western" and "non-Western" while Sjoberg [20], the pioneer of such comparisons, uses "industrial" and "pre-industrial" to denote the same two classes of cities. Despite a certain vagueness in this terminology [21], a large volume of city studies are now available for areas outside the "Western", "industrial" zone of North America, Western Europe, and parts of Oceania [22]. These and other studies [23, 24] en bloc provide a weighty volume of evidence pointing to differences in the evolution and structure of Western and non-Western cities. Already in this book the inapplicability of established theories on urban form and function have been indicated, but the differences between Western and non-Western cities have not been considered at length. Before moving forward to a detailed consideration of Kuwait's urban ecology, we must consider some of the characteristics of the non-Western city which have been noted in the literature to date.

Sjoberg [25] provides the most specific analysis available. He lists five basic differences between non-Western cities and their counterparts in Europe or North America.

1. Clustered in the central area are the "most prominent governmental and religious edifices and usually the main market". Notably, the market or commercial function is subsidiary to the politico-religious structures.

2. With a clear-cut class system, evidently the elite cling to the city centre while the poor reside at the city's periphery.

3. Further ethnic and occupational divisions in society lead to a significant degree of segregation and craft localization.

4. Land uses are not clearly distinguished since one building or plot of land can be used for several purposes (e. g. mosques serve both as schools and local market centres).

5. Finally, communication within the city and with other centres is slow and laborious—a marked contrast to the situation in industrialized cities.

Using these criteria to define the archetype "non-Western" city, we can begin to define how Kuwait differs from this traditional ideal.

2. Land Use and Urban Ecology

Sjoberg's approach to cross-cultural urban analysis is basically sociological although he refers in several instances to the association of the form and function in the urban milieu (e. g the demarcation of the centre by the markets, the square, and religious and political edifices). The concentration of these physical elements (in Kuwait, the suq, Safat Square, *Masjid as-Suq,* and the old *Baladiya al-Kuwait*—Kuwait's administrative nexus) in the centre of the Old City have already been noted. Much more important is the socio-ecological structure of the city's resident population amidst these urban landmarks.

a) Methods

In Kuwait, statistics on what in the U.S.A are called "Census tracts" are not as comprehensive as one would wish, but nevertheless the wide-ranging Census of Population of 1965 does provide a body of socio-demographic information which can usefully be employed in the analysis of the character of small districts within the city. While the population of individual Census divisions ranges from 4,650 in Shaab to 64,542 in Hawalli, the assistance of the Census Section of the Central Statistical Office made possible a further division of some of these units into smaller divisions not recognised in the Census publications.

Two analytical approaches are possible with this data. First, individual variables can be extracted and mapped using divisions such as quartile deviations. Alternatively, these variables can be grouped together and their joint distributions mapped and analysed. While the first method is simpler, it involves a strong subjective bias since the choice of the individual variables to be mapped will strongly influence the results of the analysis. While the second method is much more complex, its application has been greatly facilitated by the advent of the modern digital computer.

Both approaches are adopted here, the first to answer specific questions such as the place of residence of the rich and the poor, the citizens and the aliens; and the second to provide a wider based statistical background to substantiate much broader questions on Kuwait's structure and urban ecology.

b) Selection of variables

For the second method, using combinations of several variables, selection of the original variables was of minor importance; broadly, as many variables as possible, relating to Kuwait's ecological structure, were extracted from the Census and allied publications and reduced to percentage figures. For statistical purposes no two variables could be the reciprocals of one another (e. g., it would be impossible to use both percentage of Kuwaitis and percentage of non-Kuwaitis in the analysis).

Five individual variables were selected for mapping and analysis. The distribution of Kuwaitis and non-Kuwaitis was first chosen since this distinction underlies so many of the special characteristics of Kuwait. Then, to describe the location of the "poor" in Kuwait, two variables were chosen—percentage of illiterates and percentage employed in construction (manual labour) by districts. Since level of education largely determines one's job, salary, and hence social status in Kuwait, illiteracy was chosen as a good indication of the location of the lowest classes of society.

For the "rich", two measures of status were employed—percentage of workers in professional and technical tasks, and percentage employed in administration.

3. The Distribution of Individual Variables

a) Kuwaitis and non-Kuwaitis

Kuwaitis are strongly concentrated in the new neighbourhoods and in the outlying villages of Jalib as-Shuyukh, Idhailiya, Maqwa, and Wara, which have little or no attraction for foreign immigrants, as Text Fig. 13 shows. In contrast, over 60 per cent of the population of the Old City, of Shuwaikh, Sulaibikhat, Hawalli, and Salimiya in the suburbs, and of Dawha, Ahmadi, and

Fahahil beyond is comprised of non-Kuwaitis. The latter two centres have received an influx of foreigners because of their association with the oil industry, but Dawha is a special case. In 1961, with Iraq's invasion threat imminent, all Iraqi citizens were gathered up and enclosed within the shanty town of Dawha. Even today Dawha remains the largest concentration of Iraqis in Kuwait outside the Old City.

Within Greater Kuwait, non-Kuwaitis, denied access to the new neighbourhoods; are left to choose from Hawalli, Salimiya, Abruq Khaitan, Shuwaikh, and Sulaibikhat as places to live. In each of these districts they comprise three-fifths of the total population.

b) Illiterates

Illiteracy is strongly associated with nationality, hence the distribution of illiterates reflects the distribution of various nationalities. Iranis and Iraqis, for example, have high illiteracy rates compared with Europeans and Americans.

As Text Fig. 14 shows, these places with over 50 per cent of their male population illiterate are the outlying villages away from the main urban centres. Distance from Safat Square and illiteracy have a correlation coefficient of 0.4—significant at the 5 per cent level (females were excluded from the calculation because of

known errors in recording). Salimiya, the Kuwaiti neighbourhoods, and Ahmadi have the lowest illiteracy rates. Surprisingly, levels of illiteracy in the Old City are lower than the State average of 29.6 per cent.

c) Employees in construction

The distribution of construction workers (Text Fig. 15) reflects two trends, the first of which is the tendency of Kuwaitis not to take on manual labouring tasks. Secondly, it recognises that non-Kuwaitis constitute two major groups—those qualified to undertake "white-collar" tasks and those whose level of education restricts them to labouring.

For the first reason very few construction workers live in the Kuwaiti neighbourhoods, except those in which construction activity continues—e. g. Idhailiya and Khaldiya. There the labourers live on site in makeshift huts and shacks. Outlying centres, especially those with high proportions of illiterates (Madina al-Amal, Dawha, the desert, and Fantas) concentrate the labour force employed in construction.

From the distribution of illiterates and those employed in construction it seems that the "poor" are largely located in outlying towns and villages, although average proportions of both populations are also located in the Old City.

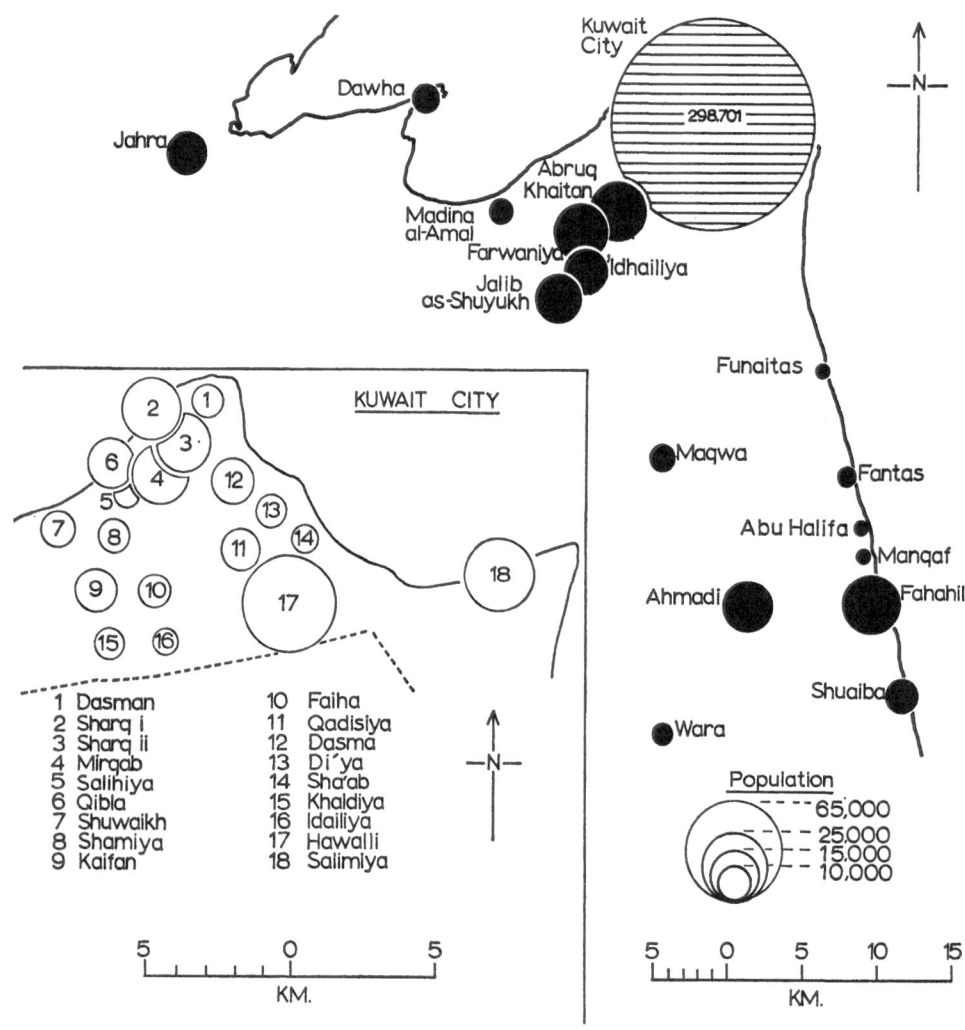

Text Fig. 12. Urban centres in 1965. Kuwait City dominates the settlement pattern (with key to districts for the Text Fig. 13—20)

d) Professional and technical workers

Since non-Kuwaitis fill most of the highly skilled positions in Kuwait, the majority of these high-grade employees live in the Old City (specifically Dasman and Salihiya, Text Fig. 16) in Hawalli, Salimiya, and Ahmadi. Less than 6 per cent of the population of the Kuwait neighbourhoods can be ranked in these classes. Figures are again lowest for the outlying centres.

e) Administrative workers

Here Kuwaitis are strongly represented, so that Text Fig. 17 shows above average proportions of administrators in the new neighbourhoods. Clearly, the preference shown for Kuwaiti citizens in the Civil Service at the Administrative level cuts across other qualifications, including educational status. This open payroll for Kuwaitis in Government offices places them higher in status than their education and other attributes would suggest (39.4 per cent of the male Kuwaitis over 10 were illiterate in 1965—Census of Population, 1965, Table 4 a). Discrimination in both housing and employment against non-Kuwaitis upsets the distribution of variables measuring both these factors. Regarding the overall structure of Kuwait City, it seems that the "rich" (the high-grade Kuwaiti employees) are being forced into the suburbs, leaving the city centre and a certain range of suburban areas open to invasion by non-Kuwaitis, themselves constituting a far from homogeneous group.

Problems arise in the separating out of individual elements for study and analysis since many are statistically associated (e. g. construction workers and male illiterates; nationality and male illiterates). Some way of sorting out groups of related variables is obviously required if we are to arrive at a more sensible account of the distribution and composition of the districts and centres comprising contemporary Kuwait.

4. The Analysis of Groups of Variables

a) Method

Psychologists were the first group of workers to face the difficulties associated with co-variance and the problem of separating meaningful character traits from a mass of information relating to psychological attributes [26]. Since then, workers in other disciplines have been confronted with the same problems [27—31]. The problem is so common that a standard solution, known as "Factor Analysis", has been evolved.

Factor Analysis is a complex statistical routine involving many steps in its calculations—so many that with a large data matrix, the routine is almost out of question unless one has the use of a computer. For the Kuwait data, a matrix of 39 observations by 38 variables was composed and prepared for the computer. A Q-mode Factor Analysis programme written by Klovan [32] was employed involving several statistical stages. Since the details of the computation are of little interest here, readers are referred to publications by Klovan [33] and Imbrie [34] for a full description.

b) Variables

Table 28 presents a full listing of the 38 variables employed in the Factor Analysis of each of Kuwait's 39 Census areas. Variables which are reciprocals of one

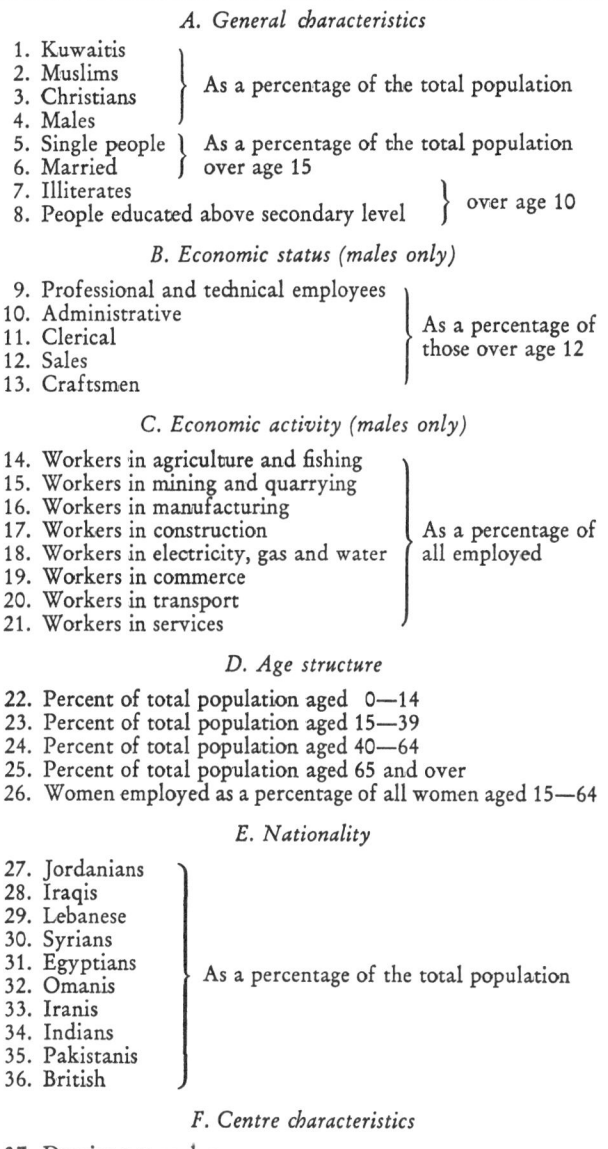

Table 28. *Variables employed in the analysis of social areas in Kuwait*

A. General characteristics

1. Kuwaitis
2. Muslims
3. Christians } As a percentage of the total population
4. Males
5. Single people } As a percentage of the total population
6. Married } over age 15
7. Illiterates
8. People educated above secondary level } over age 10

B. Economic status (males only)

9. Professional and technical employees
10. Administrative
11. Clerical } As a percentage of those over age 12
12. Sales
13. Craftsmen

C. Economic activity (males only)

14. Workers in agriculture and fishing
15. Workers in mining and quarrying
16. Workers in manufacturing
17. Workers in construction
18. Workers in electricity, gas and water } As a percentage of all employed
19. Workers in commerce
20. Workers in transport
21. Workers in services

D. Age structure

22. Percent of total population aged 0—14
23. Percent of total population aged 15—39
24. Percent of total population aged 40—64
25. Percent of total population aged 65 and over
26. Women employed as a percentage of all women aged 15—64

E. Nationality

27. Jordanians
28. Iraqis
29. Lebanese
30. Syrians
31. Egyptians
32. Omanis } As a percentage of the total population
33. Iranis
34. Indians
35. Pakistanis
36. British

F. Centre characteristics

37. Density per sq. km
38. Size of centre in thousands

another (e. g. Kuwaitis and non-Kuwaitis; people over 15 and people under 15) must be avoided since they will be perfectly correlated. All variables were transformed into percentage distributions to obtain normal or near-normal distributions in every instance. All 39 Census areas in Kuwait have been employed in the analysis to give a full cover of the extent of the socio-economic gradients throughout the State.

5. The Factors: Their Composition and Distribution

Broadly, the programme "collapses" the original 39×38 data matrix into a series of "Factors". These Factors are a synthesis of the variance of the original variables produced from a cos-theta matrix, itself the basis for the calculation of eigenvalues and eigenvectors used to construct the principal component factor matrix.

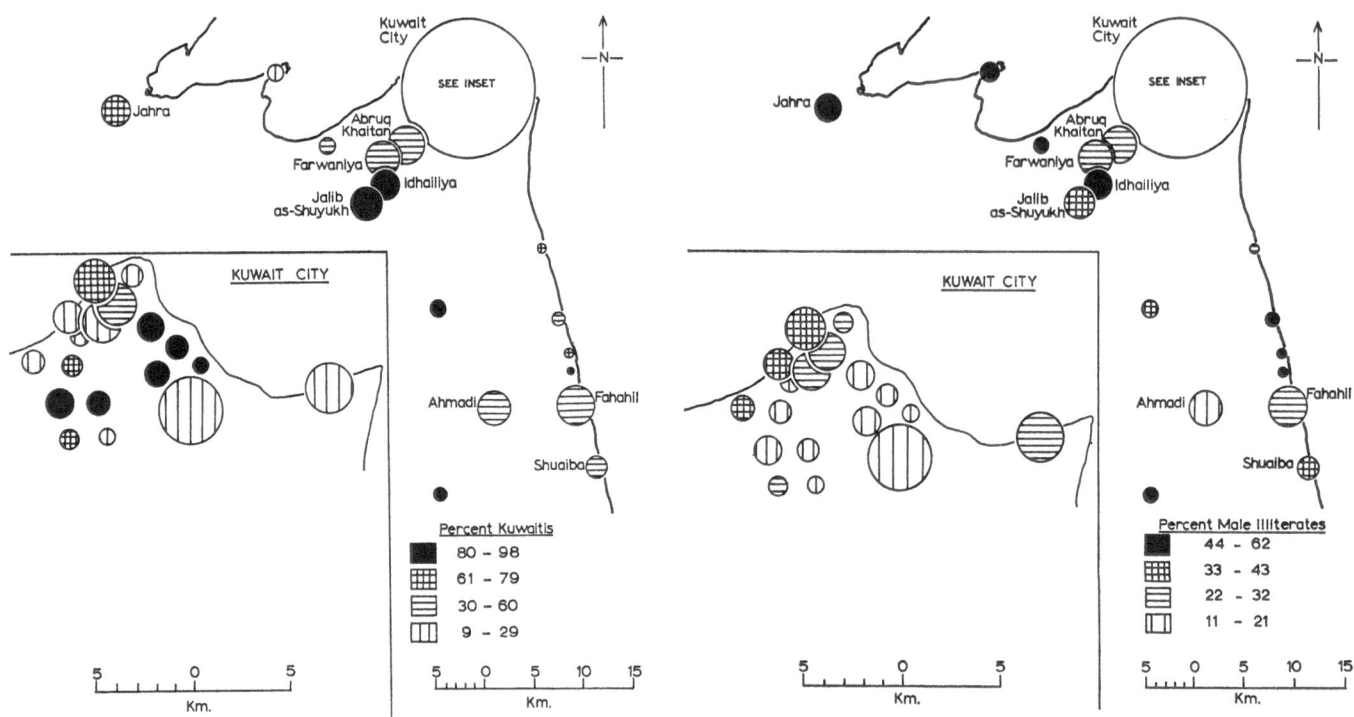

Text Fig. 13. Distribution of Kuwaitis in 1965.
(For key to districts see Text Fig. 12)

Text Fig. 14. Distribution of illiterates in 1965.
(For key to districts see Text Fig. 12)

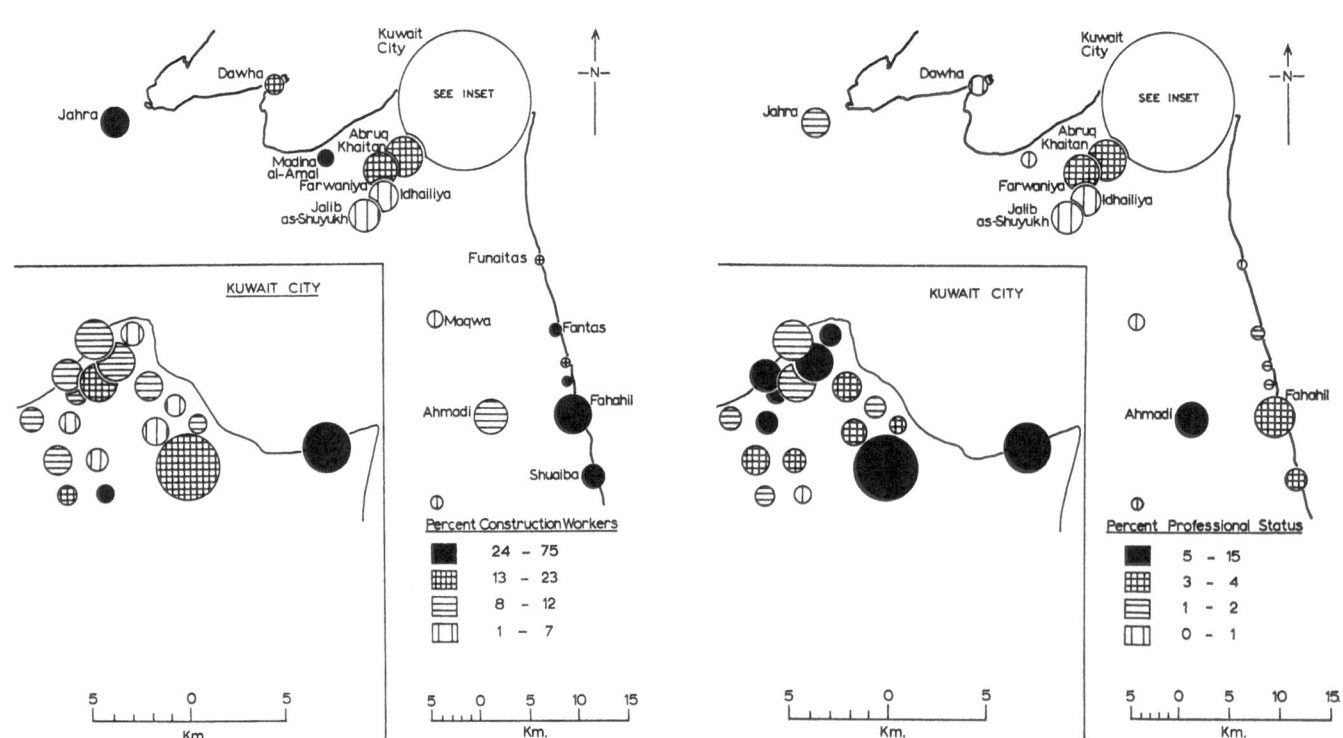

Text Fig. 15. Distribution of construction workers in 1965.
(For key to districts see Text Fig. 12)

Text Fig. 16. Distribution of professional and technical workers
in 1965. (For key to districts see Text Fig. 12)

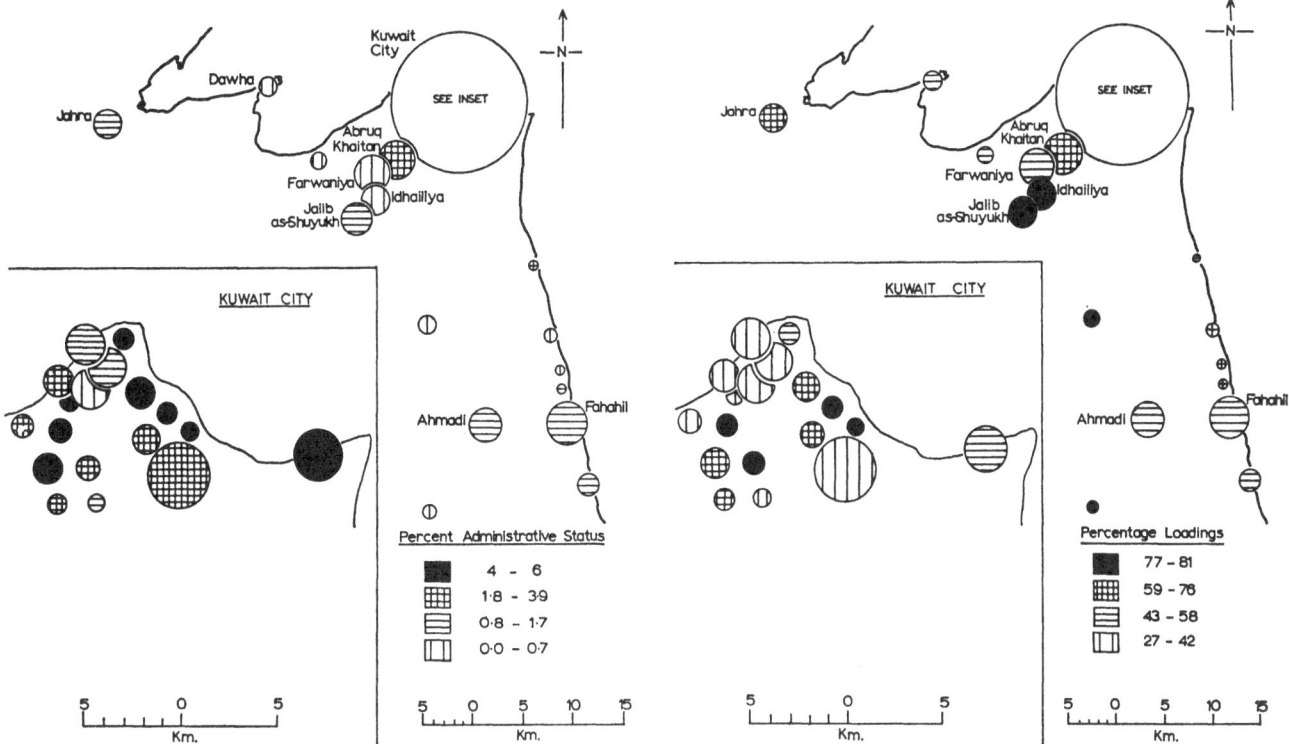

Text Fig. 17. Distribution of administrative workers in 1965.
(For key to districts see Text Fig. 12)

Text Fig. 18. Distribution of Factor 1.
(For key to districts see Text Fig. 12)

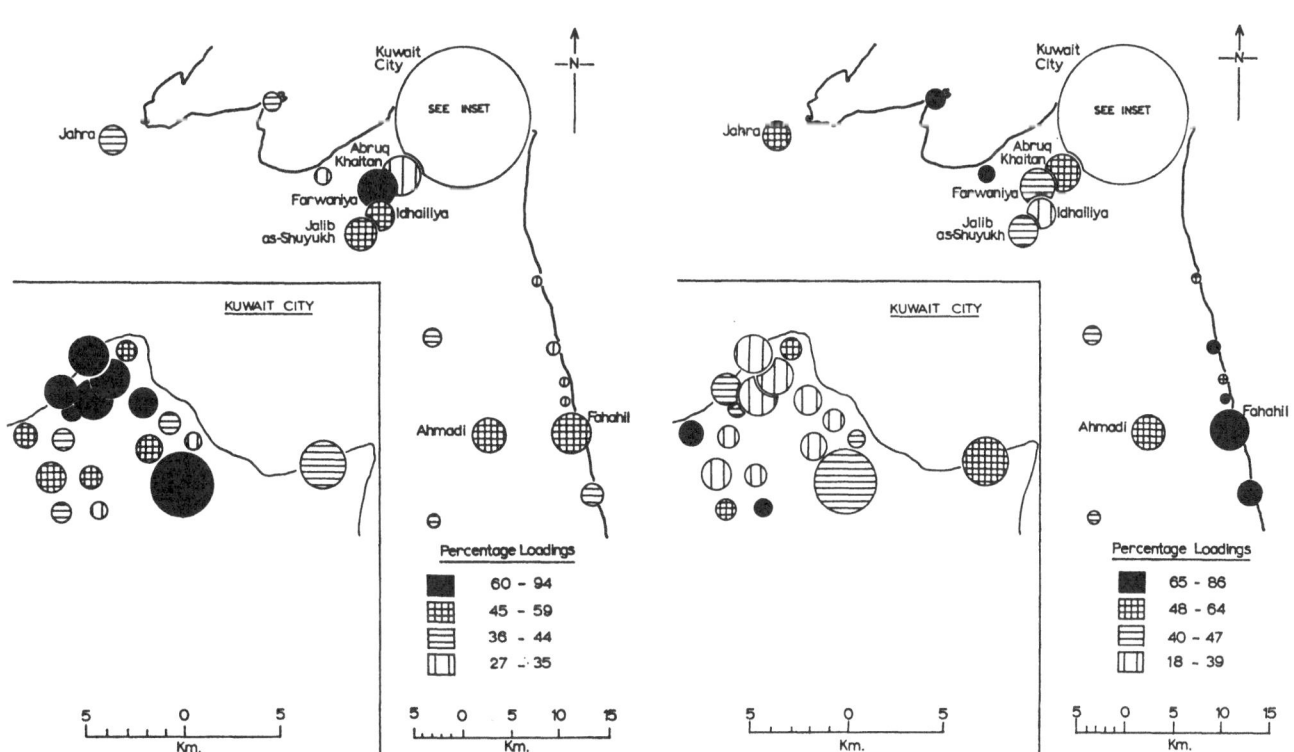

Text Fig. 19. Distribution of Factor 2.
(For key to districts see Text Fig. 12)

Text Fig. 20. Distribution of Factor 3.
(For key to districts see Text Fig. 12)

Once these synthetic "Factors" have been evolved, each of the 38 original variables is regressed in turn on the Factors so that the composition of each of the latter can be ascertained. Secondly, each area is attributed weightings on each of the Factors (using a "Varimax Factor Matrix") so that the areal distribution of the Factors can be plotted on maps. Finally, the varimax matrix is converted to Factor Components for the last 3 Factors to facilitate the plotting and identification of similar areas on 3-component diagrams.

Each of these 3 analytical steps will be treated separately in the case of Kuwait.

a) Factor composition

Using the 38 variables specified for all 39 Census areas in Kuwait, this large data matrix can be satisfactorily condensed into 7 Factors which together "explain" over 99 per cent of the variation in this matrix. However, the first 3 Factors "explain" 92.0 per cent of the variance while the subsequent 4 Factors only account for an extra 7.1 per cent of the total explanation. We are justified in considering only these first three Factors since they account for the vast majority of the variance of the original data matrix. Each of the three Factors expresses one unique dimension of the original matrix and hence is comprised of a different group of elements from all others.

Factor 1. The first Factor "explains" 38.7 per cent of the total variance. Factor 1 is comprised of a variety of elements, but the variables with high positive or high negative scores are listed in Table 29. Clearly, most of the attributes outlined are those of the Kuwaiti citizen population, and indeed this variable scores most highly on Factor 1. As shown above, Kuwaitis are a very young population with a high proportion of married people; they work largely in service industries, and are almost all Muslims. Conversely, they are neither craftsmen nor construction workers—both attributes of non-Kuwaitis generally and Iranis in particular.

Table 29. *The composition of Factor 1*

High positive loading		High negative loading	
1. Kuwaitis	4.09	13. Craftsmen	1.13
21. Service employment	2.60	17. Construction	
2. Muslims	2.17	workers	0.89
7. Illiterates	1.59	27. Jordanians	0.65
22. Age 0—14	1.45	33. Iranis	0.52
6. Married	1.12	38. Size of Centre	0.25

Factor 1 fairly summarises the attributes of the Kuwaiti population as described in previous Chapters; for this reason we can name this dimension the "Kuwaiti citizen" Factor.

Factor 2. As Table 30 outlines, the populations described by Factor 2 live at a high density, are Muslims, have a male bias, and are concentrated in the young active age groups (15—39). They tend to work in commerce or be craftsmen, but they are not agricultural or construction workers, neither are they Kuwaitis.

We can conclude that Factor 2 is describing a section of the non-Kuwaiti population. From their occupation (craftsmen and employed in commerce but not manual

Table 30. *The composition of Factor 2*

High positive loading		High negative loading	
37. Density	5.60	1. Kuwaitis	0.48
4. Males	1.07	14. Agriculture and	
2. Muslims	0.98	fishing	0.13
23. Age 15—39	0.93		
38. Size of centre	0.86		
6. Married	0.77		
13. Craftsmen	0.72		

labourers), it appears that they are a better educated section of the immigrant community containing a preponderance of males and of married people. These two variables are not irreconcilable since many married non-Kuwaitis are resident in Kuwait but without their wives. Hence we can call this dimension the "Higher status non-Kuwaiti" Factor.

Factor 3. The attributes of the populations described by Factor 3 are almost the inverse of the second Factor (Table 31). Factor 3 indicates that the third dimension of

Table 31. *The composition of Factor 3*

High positive loading		High negative loading	
17. Construction		37. Density	1.59
workers	2.82	21. Service employment	1.14
13. Craftsmen	2.70	11. Clerks	0.33
7. Illiterates	2.07	19. Commerce	0.33
2. Muslims	2.05	12. Sales	0.24
4. Males	1.93	38. Size of centre	0.18
23. Age 15—39	1.60		
33. Iranis	1.12		

variance in Kuwait is that containing illiterate, male, Muslim, construction workers or craftsmen with a large proportion of their total population in the young active age groups. Significantly, a further characteristic is Irani nationality. These people do not live in large centres and do not have jobs as clerks or in commerce. This third dimension, explaining 28.1 per cent of the overall variance, can be titled the "lower status non-Kuwaiti" Factor.

While these 3 Factors account for 92.0 per cent of the total variance, *Factors 4 and 5* (explaining only 4.2 and 1.5 per cent respectively) are in fact sub-sets of Factors 2 and 3. Factor 4 describes a population of medium status, employed in services but not in construction, containing single people and a male bias. The Factor Score Matrix associates these characteristics with Jordanians.

Factor 5 describes a married population employed in mining, containing young people under 15, and largely Christian. They are people of professional and technical status together with some clerks. Their nationality is Pakistani or Indian. This dimension is describing the characteristics of Ahmadi.

As the foregoing suggests, the analysis proceeds by selecting out the most salient dimensions of variance first of all, proceeding steadily towards the description of smaller and smaller dimensions, sometimes explaining the attributes of only one centre (e. g. Factor 5 with Ahmadi). Our analysis of these Factors cannot proceed indefinitely, so that in the subsequent section only the distribution of Factors 1, 2, and 3 will be discussed, since they contain 92.0 per cent of the descriptive power of the original 38 variables.

b) The distribution of factors throughout Kuwait

Plainly, our major interest lies in the distribution of Factors 1, 2, and 3 throughout Kuwait. Two methods of analysis are employed; first, each of the three Factors is plotted separately on a map of the Census areas of Kuwait and its distribution represented choroplethically. Secondly, groups of similar areas are identified using 3-component graph paper.

6. The Geographic Distribution of Individual Factors

Factor 1. Since Factor 1 largely describes the Kuwaitis and their attributes, the areas with the highest positive loadings are the Kuwaiti suburbs of Kuwait City (key numbers 9—15), together with the outlying villages of south Kuwait and the east coast. Districts with high negative loadings are conversely those areas most touched by immigration (e. g. the Old City, Ahmadi, and Fahahil). Text Fig. 18 illustrates the distribution of Factor 1.

Factor 2. This Factor, named the "Higher Status non-Kuwaiti" component, selects out those areas of non-Kuwaiti settlement which contain a large proportion of their labour force in craft jobs (e. g. commerce) but with few manual labourers (Text Fig. 19). Hawalli, Farwaniya, and Sharq 1 and 2 with Mirqab in the Old City score highly on this component. Two groups of areas have very low scores on this axis—those districts with a strong Kuwaiti bias (see above), and those districts containing low status manual labourers such as Madina al-Amal and Shuaiba.

Factor 3. It is this latter group which the third component specifies. It includes areas under construction (Idailiya and Salimiya) as well as industrial areas (such as Shuaiba and Shuwaikh), both of which are associated with non-Kuwaitis of working age and of low educational standing (Text Fig. 20).

7. Combinations of all 3 Factors

The programme, which provides row-normalized Varimax Factor Components, allows us to plot the distribution of each one of Kuwait's 39 Census areas on triangular co-ordinate graph paper. This diagram (Text Fig. 21) arrays all the 39 areas on the three Factor axes simultaneously, so that groups of areas with similar attributes can be readily distinguished. Four major groupings can be distinguished on the diagram which incorporates 92.0 per cent of the descriptive power of the original 39×38 data matrix.

a) Group 1: 10 areas containing higher status non-Kuwaitis

Criteria

1. *"Kuwaiti Factor":* Loadings under 35 per cent
2. *"Higher status non-Kuwaiti Factor":* Loadings over 30 per cent

3. *"Lower status non-Kuwaiti Factor":* Loadings under 50 per cent

These areas are the districts most affected by immigration and consist of all the areas of the Old City, the oil towns of Ahmadi and Fahahil, and other suburbs where immigrants concentrate in considerable numbers. In every instance, more than 60 per cent of the total district population consisted of non-Kuwaitis. The immigrants are employed in commerce and skilled and semi-skilled occupations however, and very few are manual labourers. Factor 2 fairly summarizes the attributes of the people resident in these areas.

b) Group 2: 11 areas containing mostly lower status non-Kuwaitis

Criteria

1. *"Kuwaiti Factor":* Loadings under 40 per cent
2. *"Higher status non-Kuwaiti Factor":* Loadings under 30 per cent
3. *"Lower status non-Kuwaiti Factor":* Loadings over 45 per cent

As the choroplethic map for Factor 3 showed (Text Fig. 20), this group of areas is composed of small outlying villages (Dawha, Manqaf, Fantas, Madina al-Amal) together with new areas under construction (Salimiya, Abruq Khaitan, Sulaibikhat, and Idailiya). As indicated above, building labourers live on site and are hence the dominant demographic groups recorded in the Census for these areas. Idailiya, a new Kuwaiti suburb, has the most extreme loadings on Factors 1 and 3 for this reason.

c) Group 3: 6 areas containing higher proportions of Kuwaitis and some lower status non-Kuwaitis

Criteria

1. *"Kuwaiti Factor":* Loadings over 45 per cent and under 65 per cent
2. *"Higher status non-Kuwaiti Factor":* Loadings under 20 per cent
3. *"Lower status non-Kuwaiti Factor":* Loadings under 30 per cent and under 50 per cent

Excluding Khaldiya, a new Kuwaiti suburb in course of construction in 1965, all the areas in this group lie beyond the built-up area of Kuwait City. Most of the districts are old-established Kuwaiti villages (Jahra, Failaka, Abu Halifa, and Funaitis) which remain over 70 per cent Kuwaiti in composition, sustaining only slight intrusions of non-Kuwaiti immigrants and most of these associated with the construction industry. Between 22 and 54 per cent of the population in all 6 areas can neither read nor write.

This group is not strongly loaded on any one factor; it does display a low negative loading on the "Higher Status non-Kuwaiti" factor so that the areas included in the group are represented by the distribution of the bottom quartile of Factor 2 (Text Fig. 19).

d) Group 4: 12 areas containing Kuwaitis and a smaller proportion of higher status non-Kuwaitis

Criteria

1. *"Kuwaiti Factor":* Loadings over 45 per cent
2. *"Higher status non-Kuwaiti Factor":* Loadings under 40 per cent

3. *"Lower status non-Kuwaiti Factor"*: Loadings
   under 25 per cent

This group of 12 areas consists largely of strongly
Kuwaiti areas. We can subdivide the areas into 2 sub-
groups on the basis of Factor 2, using the threshold of
20 per cent as the dividing line.

Group 4a. Most of this sub-group consists of the
Kuwaiti suburbs of Shamiya, Kaifan, Faiha, Qadisiya, and

Group 4b. By comparison, the 5 areas in this sub-
group (Di'ya, Shaab, Shadadia, Maqwa, and Wara) are
more exclusively Kuwaiti. Di'ya and Shaab are new
neighbourhoods in the suburbs of Kuwait City which in
1965 were still under construction (viz. the loadings on
Factor 3 are higher than for Group 4a). Shadadia,
Maqwa, and Wara, by contrast, are largely shanty camps
where Kuwaiti Badu live. Shadadia is situated just south

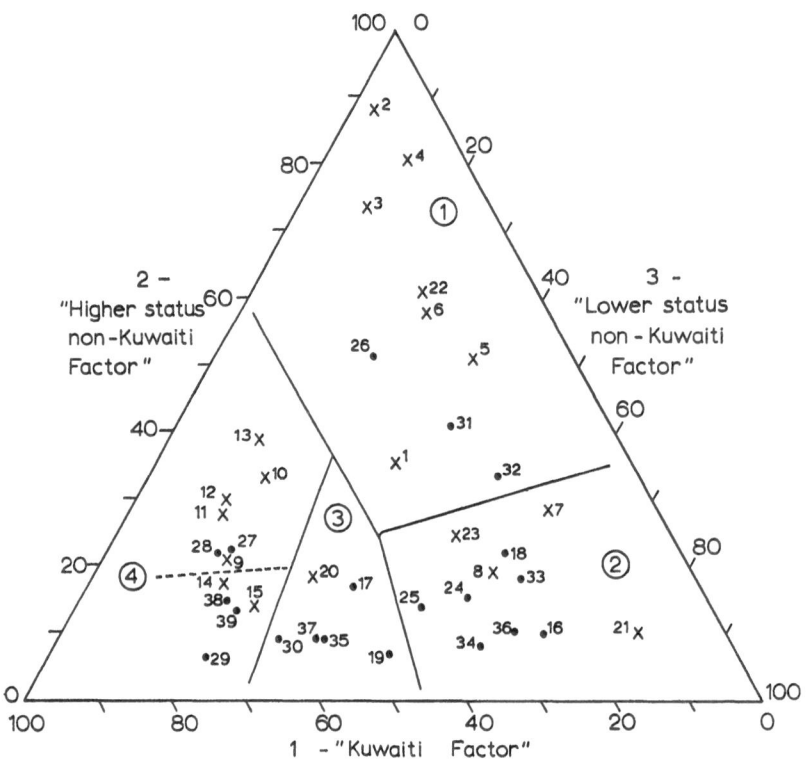

Text Fig. 21. Three co-ordinate graph for Factors 1, 2 and 3

Key: 1 Dasman   15 Sha'ab   27 Jalib
      2 Sharq 1   16 Madina   As-Shuyukh
      3 Sharq 2   al-Amal   28 Idhailiya
      4 Mirqab   17 Jahra   29 Shadadia
      5 Salihiya   18 Dawha   30 Failaka
      6 Qibla   19 Desert   31 Ahmadi
      7 Shuwaikh   20 Khaldiya   32 Fahahil
      8 Sulaibikhat   21 Idailiya   33 Shuaiba
      9 Shamiya   22 Hawalli   34 Manqaf
     10 Kaifan   23 Salimiya   35 Abu Halifa
     11 Faiha   24 Jabriya   36 Fantas
     12 Qadisiya   25 Abruq   37 Funaitas
     13 Dasma   Khaitan   38 Maqwa
     14 Di'ya   26 Farwaniya   39 Wara

Dasma, together with the new centres of Jalib as-Shuyukh
and Idhailiya situated just 3—5 km south of Kuwait
City. These suburban areas by 1965 were fairly well
established. Scores on Factor 2 are between 20 and 40
per cent because of the numbers of non-Kuwaiti house-
boys and housemaids (mostly Omanis and Iranis) resident
with Kuwaiti families in these areas. These workers employed in domestic services, and sometimes living as a
family within a Kuwaiti villa, account for the higher
than expected scores of these areas on the "Higher Status
non-Kuwaiti" Factor.

of Kuwait City but Maqwa and Wara are located on the
Burqan oilfield.

It may be that since part of Factor 1 is made up of
illiterates that Groups 4a and 4b (from the Kuwait
suburbs of Kuwait City to the shanty towns of Maqwa
and Wara) are arrayed along a literacy gradient. In
Maqwa and Wara, over 40 per cent of the males over
age 10 could neither read nor write, while an average
figure for these suburbs was only 20 per cent. Hence, the
three-component diagram has added a further dimension
to the 3 Factors already indicated. This dimension is

apparently reflecting a trend towards higher literacy amongst Kuwaitis in the suburbs. Since literacy is associated with other indicators of advancement (e. g. economic activity and status of employment), we can call this gradient recognized amongst Kuwaiti areas a "modernization" axis.

8. Factor Analysis and the Structure of Kuwait City

Having indicated the dimensions of socio-economic variance throughout Kuwait, we can return again to the examination of the detailed structure of Kuwait City. Greater Kuwait, the built-up area within the fourth ring road, comprises 19 Census tracts, all of which have been employed in the foregoing factor analysis. A distinctive symbol has been used on the 3-component graph (Text Fig. 21) to bring out those districts within Greater Kuwait.

The diagram shows that the 19 areas composing Kuwait City are widely dispersed on all 3 Factors. Indeed, within Kuwait City we are dealing with a degree of social and economic heterogeneity at least as large in the State as a whole. Conversely, if we consider only the centres other than those included in Kuwait City, Text Fig. 21 shows that most of their variance can be represented on Factors 1 and 3 without reference to Factor 2. In other words, the dimension of Factor 2, "Higher status non-Kuwaitis", is largely restricted to Kuwait City. There are three notable exceptions to this generalization—Ahmadi, Fahahil, and Farwaniya. The latter is a special case since it acts as a non-Kuwaiti dormitory town for Kuwait City, while the other two are directly involved with the oil industry. Hence all three contain a sizeable proportion of "Higher status non-Kuwaitis".

Considering the distribution of the districts of Kuwait City on all three axes simultaneously, we can use the same four divisions employed in the analysis of the Factors for the whole State.

9. Social Areas in Kuwait City

While Text Figs. 18—20 show the distribution of areas shaded for single Factors, our concern is to group the district of Kuwait City into sets of similar and dissimilar areas.

a) The Old City and other areas of immigrant invasion

The same criteria as were used to define Group 1 above again apply. Within the Group (the Old City with Shuwaikh and Hawalli) considerable variation is recorded, particularly on Factor 3. Dasman is clearly the most cosmopolitan area in Kuwait with almost equal loadings on all 3 Factors. Sharq 1 and 2, and Mirqab, contain a heavy loading on the higher status immigrant axis and low loadings on the lower status immigrant axis. New housing developments—both flats and bungalows—are attracting, for example, Britons and Americans, Indians, Pakistans, and Jordanians in sizeable numbers, all of whom have a relatively high level of education and social status. In 1965 these nationalities together numbered 26,660 in the Old City of a non-Kuwaiti total of 70,340 (Census of Population 1965, Table 2).

Interestingly, the analysis closely associates Salihiya and Qibla with Hawalli. All three areas score under 20 per cent on Factor 1, medium-high loadings on Factor 2, and low loadings (under 40 per cent) on Factor 3. From Tables 29—31 showing Factor composition, it seems that these areas contain a variety of people, but in the main married non-Kuwaiti Muslims with medium status jobs—possibly Jordanians.

Finally, the widespread construction activity in Dasman (including the building of a Hilton Hotel) involves manual labour, thus raising the area's loading on Factor 3.

b) Areas under construction with low-status immigrants

As indicated previously, Factor 3 consists primarily of illiterate male craftsmen construction workers, probably Iranis. Sulaibikhat, Idailiya, and Salimiya were all in course of construction in 1965 and thus score highly on this Factor. With Shuwaikh, the explanation differs slightly since the area contains the docks, several factories, and some small workshops. In all three tasks, male manual labour is required.

c) Strongly Kuwaiti areas

Factor analysis reveals that the new Kuwaiti neighbourhoods (Shamiya, Kaifan, Faiha, Qadisiya, Dasma, Di'ya, Shaab, and Khaldiya) stand out clearly as the most distinctive group of areas in the city, much as one would expect in view of the policy of discrimination against non-Kuwaitis. Some invasion of the oldest inner suburbs (e. g. Shamiya and Kaifan) by higher status non-Kuwaitis is taking place, probably in association with domestic services (see above) and the new co-operative shopping centres. Khaldiya is quite heavily loaded on Factor 3 because both housing developments and the buildings for Kuwait University were under construction in the district at the time of the Census.

Equipped with this statistical information on the social areas of Kuwait City, we can return to the issues raised at the outset of this Chapter concerning city structure and the location of the rich and the poor classes of society. If Kuwait is to parallel the idealized model suggested for non-Western cities, as suggested by Berry [35] and Sjoberg [36], then we can expect high-status areas to be located in or near the city centre with a downward gradation in social class towards the periphery.

10. International Parallels

This study of the social areas of Kuwait draws attention to the limitation of Western-derived theory to describe the urban ecology of an oriental city. However, in the Orient as a whole, several case studies are available which indicate that the structure of Kuwait's built-up areas has parallels elsewhere in the East.

Breese [37] in a survey entitled "Urbanization in Newly Developing Countries" has an important section on Indian cities, particularly Delhi. There, an influx of refugees after partition produced a characteristic grouping of squatters' areas located on open spaces near the city's edge. In addition, high class suburban housing is being added at random to the periphery of most larger

Indian cities, providing a stark contrast to the over-crowded and deteriorating Old City. In a sense Kuwait possesses the same three areas—shanty towns (e. g. Maqwa and Wara), new suburbs, and the Old City—although the quality of buildings and social amenities in all three areas is much higher in Kuwait than in other developing nations of Asia.

Again from India, Brush [38] cites several examples of dual urban development arising from culture contact—specifically, the British period. Subsequent planned developments (e. g. at Modinagar, Jamshedpur, and Chandigarh) are resulting in the emergence of self-contained cells with one land use per block. Both dualistic development and strongly segregated land uses are characteristics of the Kuwait urban area.

From South-East Asia, McGee indicates that dualism is a characteristic of both economic activities ("bazaar-type" and "firm-type" economies) and residential development. He writes:

"The major element of the colonial city was the mosaic of ethnic quarters—the tightly packed shop-house areas of the Chinese, the spacious low density 'compounds' of the Euro-peans, and the rural-like villages of the indigenous population scattered around the fringes of the city. The rapid growth of the population of the cities in the post-war era, associated with the socioeconomic changes which are creating an emergent middle class, have caused a proliferation of squatter resettle-ments in the interstices and fringes of the city, as well as the growth of western-type suburbs, adding new elements to the residential ecology of the city. In the process some of the lines between the various racial enclaves have become a little blurred, but overall ethnic concentration is still responsible for the major divisions in the residential areas of the city." [39]

Thus Kuwait's urban ecology, as brought to light by factor analysis, is by no means exceptional in the context of non-Western cities in that racial and national group-ings are the elements most strongly underlying the city's

structure. While Western-derived theories relating to the ecological structure of cities in the Orient proves in-adequate to explain the structure of Kuwait City, it seems that there are several close parallels of the situation in Kuwait from at least two major regions of Asia.

11. Conclusion

At several stages of this analysis of Kuwait's urban development reference has been made to the dissimilarity between urbanization in Western and non-Western areas. In this Chapter it seems we have a strong statistical base for drawing conclusions on Kuwait's urban structure. Despite exceptional trends, such as the evacuation of Kuwaiti citizens from the Old City into the suburbs, Kuwait's experience closely parallels that of several widely scattered Asian countries. Factor analysis revealed that within Kuwait City, quarters containing a majority of Kuwaitis had such different socio-economic charac-teristics from quarters containing a majority of non-Kuwaitis that the two districts were incomparable using the three major components of variance extracted from the original 39×38 data matrix. Increasingly, we are led to the conclusion that cities of the Third World generally—with Kuwait among them—possess attributes and characteristics peculiar to themselves.

The description and analysis of the population of Kuwait using economic, social and demographic charac-teristics drawn largely from the census has its limitations, for it is only through a full understanding of the total ecology of the people caught up in the urbanization process that its implications at all levels can be fully appreciated. The following chapters elaborate more fully on the pattern of health and illness among the hetero-geneous groups comprising the population of Kuwait.

VII. Health and Disease

1. Introduction

Sweeping statements concerning the evil effects of urbanization are no more justified than the opposing view that urbanization ultimately leads to Utopia. A variety of incentives such as the degree of industrial development, anticipated increased earning power, the fading quality of the traditional pastoral or nomadic life and a host of other factors are all concerned in shap-ing the new environment and the individual's reaction to it. But if we are going to take the rational view that urbanization should be the qualified and controlled immi-gration into and around an urban nucleus, we must distinguish it from "pseudo-urbanization" which is the exact antithesis, with the all too familiar shanty-towns radiating from or encircling the original organised settle-ment. For it is this hybrid form of urbanization which possesses all the potential for a rapid decline into the worst of slum conditions which breed mental, moral and physical degradation.

a) Pseudo-urbanization

In Kuwait, over the recent past, pseudo-urbanization has been a minor problem, and by far the most significant

evolution has been orderly within organised and planned townships. While long-continued pseudo-urbanization must obviously be condemned as socially and medically undesirable, in the short term it has perhaps the one merit in that it allows the impact of the changing pattern of life to be softened, particularly for the older people, while at the same time certain new experiences such as children's education and family medical services, when and if they can be provided, can begin to alleviate the deficiencies and afflictions of their former life. This is particularly true of those affecting women in the child-bearing age group and the children. Such an event in Kuwait, and indeed elsewhere, when it has been con-trolled, has served to soften abrupt transition from the freedom and austerity of life in the desert to the apparent restrictions yet intemperance of city dwelling. To give but two examples of the problem. Human defaecation among the desert Badu is performed in the sand outside the tent (Fig. 38). Sun and wind combine to desiccate the stool removing any offensiveness and, more important, render it a poor breeding ground for flies. The evidence of this simple method of relieving the bowels can be found in any recently deserted encampment for the dor-mant grass rapidly responds to the unexpected stimulus of moisture, shade and fertiliser. The second example is the Badu desert custom of burying their dead in a pit very near the tent. Neither of these practices are acceptable in a shanty (serifa) area and a change is soon forced upon the new arrivals. It should be remembered that as soon as durable walls of houses are built, areas of permanent shade are created: these can be readily fouled and become damp with the inevitable fly-breeding a serious nuisance and hazard. This sort of thing can be a frequent occurrence if there is no transition phase, which should be used by the authorities, if they have the facilities, as a period of education and adaptation. Public education of this problem in all sectors of the population can assist in containing it.

b) The health of the Badu

It is salutary to remind ourselves that the Badu have long had the reputation for enjoying a resilient constitution and if death in infancy and early childhood can be averted they may survive for many years, remaining remarkably free from disease and in particular harbouring few intestinal parasites, to whose transmission cycle the desert presents a very unsympathetic attitude. But if the Badu enter an urban community their comparative inexperience of infectious disease can rebound on them, the way being open to serious and rapid infection against which they may have little resistance. Later we will examine these effects along with the more nebulous consequences of urbanization which give rise to those common "blessings" of the modern way of life; dental caries, diabetes, obesity, hypertension and other non-infective conditions, either within or beyond the control of the individual.

The justification for such an approach is that it must follow that when future changes in the social patterns of human lives can be anticipated, the environmental effects upon these should also be foreseen and steps taken to educate people to control the changes or themselves in order to live and enjoy a satisfying, happy and productive life; beyond this point the present authors do not feel competent to prescribe.

c) Effects of urbanization

The continuation of urbanization as a feature of man's progress is inevitable and there is sufficient evidence that the strain it is putting upon his capacity for adaptation is creating ill-effects on his mental and physical health, analogous to those arising from the Industrial Revolution in Europe in the eighteenth and nineteenth centuries. In a small country this is immediately noticeable whereas the large conurbations of a big country such as India have only a minor impact upon that country's problems as a whole. The results of these stresses are not always immediately apparent. They may be identified by uncontrolled changes in the cultural behaviour of large numbers of people which then becomes the concern of the guardians and students of public morality and conduct, be they religious, sociological, recreational or legal. These ill-effects may rapidly spread into the fields or art, architecture and those sections of industry and commerce which exist by flattering and indulging the current tastes. Very soon those people in the public services, principally the medical and social workers, who are intimately concerned with individuals finding the going of adjust-

ment too hard, often come to look on the changing scene with a somewhat jaundiced eye, seeing not the wood of success for the trees of imperfection.

d) Integration of planning for social and environmental changes

An appreciation of human biology and ecology today demands a cross-fertilisation of knowledge and ideas; a sharing and understanding of other peoples' interests, and a training for responsibilities which many professional workers have not had the time to gather during a specialised education. That this is true for the medical profession has recently been recognised in Britain by the publication of the Royal Commission Report [1] which emphasises the importance of a more selective and vocational approach to a young doctor's training.

e) The importance of integrating disciplines beyond fringe of medicine

Man's immediate environment can expand so quickly that there is no protection by distance, oceans, mountains or skies against exotic germs or ideas [2]. To anticipate change with the object of at least partially controlling it without necessarily inhibiting it, demands accumulation of knowledge by human biologists, be they doctors or others, from varied sources, many of which still play no part in their education. Clinical disease and epidemiology, public health engineering, veterinary medicine and microbiology, adaptation physiology of man and animals, genetics, social athropology, land-use, soil chemistry, geology, climatology and economic geography, all of these are essential. But it is not enough to know what to do; we must know why. We are now faced with a daunting array of moral problems in our daily lives both as lay people and as physicians. To withstand and grapple with them demands a wide understanding of those factors which govern man's behaviour and too often lead to his fall.

Man and his animals may enter a closed or restricted environment and upset a previously stable ecology; recent examples have been the development of vast inland water systems designed to increase irrigation of barren land and develop power for industrial purposes.

f) Adaptation of the individual to meet the new stresses

We see in our own cities major rehousing schemes when families have been lifted many feet into the air and must learn to develop new patterns of social intercourse. The effects of these have been to inject sudden new and unrecognised stresses into the lives of people by disrupting their social customs and creating new habitats for disease-bearing vectors which are able to attack a population which has had no time to acquire immunity. It seems that time is all important to allow for the development of this immunity whether it be against infectious disease or the distorted notions brought about by modern urbanization and industrialisation. Jacques May [3] has expressed this lucidly: "The environment should be considered as a reservoir of challenging stimuli, acting on hosts endowed with a certain genetic make-up and certain characteristics acquired as a result of their experience in the environment in which they have grown (such as the scars of immunising encounters). The resulting disease pattern is an expression of temporary maladjustment of the host to the set of challenges he is confronted with."

g) Alterations in disease patterns

The importance of recognising that post-exploratory penetration, migration and redevelopment can bring with them new, or alterations in the older patterns of disease, must be emphasised. Pavlovsky [4] has defined what he called a "landscape epidemiology", relating terrain to disease, endemic or enzootic and Hoare [5] has illustrated this: "Thus in desert areas of Asia and North Africa, inhabited by burrowing rodents, the presence of oriental sore might be suspected: in thickets among rock formations of America containing armadillos, one might expect to encounter vectors of Chagas' disease (South American trypanosomiasis), while in tropical Africa, woodland with big game or river banks with thick vegetation harbouring tsetse flies might indicate the presence of sleeping sickness. If the potential danger of such places is recognised, they can either be avoided, or appropriate measures can be taken to protect human immigrants from infection." It is not difficult to enlarge this to include similar habitats in an urban context be they sites of infection or, more frequently neurotic disease.

h) Definition of medical geography

Medical geography is the knowledge of the distribution, ecology, habits and diseases of communities in all parts of the world. For many years this has been taken to apply only to the tropics, but penetration of man into cold regions and over-crowding in temperate regions has broadened the concept and pointed the need for a universal application. This requires sources of information not easily available and whose applicability may not have general validity. Maegraith [6] has attempted this in relation to the distribution of disease, and recent German publications (Jusatz [7, 8]) have begun to clarify the picture in individual countries. Others have accomplished the wider task of examining the relation of the patient and his specific disease in his environment [9, 10], which will be our endeavour with regard to the wider field of a national pattern of disease.

The majority of infectious diseases are at present preventable, containable or curable, but many of the disorders which have come to plague us are non-infectious, and are the result of a breakdown of natural resistance, whether it affects the body or the mind. Some of these derive directly or indirectly from the hand of man himself and even his physician! The hazard of the opportunistic disease is upon us often following the suppression of normal immunity by natural disease or its treatment. The "diseases of medical progress" are causing increasing concern for they are created by the new diagnostic and therapeutic tools in the hands of doctors and technicians. Now the concept of 'diseases of technical progress' is recognised, applying to conditions arising, though often remote from, the original source, through methods of manufacture, packaging and transport which may increase the shelf-life of food and other products and inadvertently alter the micro-climate causing toxic substances to develop or be introduced.

The late Professor Dudley Stamp [11], in the Heath Clark lectures for 1964, pleaded for the use of the map as a tool for the doctor as well as the geographer. Stamp was critical of the narrow outlook of the modern professional man on his graduation when he said "we know more about the effects of local differences of climate on the life and well-being of fruit trees than we do of their effect upon the life and well-being of human beings". He pointed out the ignorance which people demonstrate when siting and building houses without considering the effects of water drainage, air movement and drainage, shade and insolation and the proper use of landscaping. Prevalence records of disease usually show an intensity in poorer housing areas and it can be assumed that this type of housing is usually developed on sites which are neither as attractive nor as salubrious as privately owned and developed properties. Overcrowding and poor siting of dwellings inevitably leads to disease even in comparatively satisfactory accommodation. Tropical disease is no more than the result of poor personal and public hygiene in an environment which predisposes to the breeding of vectors and in which occur the natural hosts of human and animal infection.

Siting of industrial complexes is of vital importance now that we have evidence of what damage air pollution can do. To quote Stamp once again, "Danger lies not in the average or mean conditions but in the exceptional, and especially in the rare combination of exceptional circumstances which may occur but once in a century".

A final thought before passing from the general to the specific problems peculiar to Kuwait. For many years certain substances have been known to be essential for vital processes to operate effectively: these elements belong either to the electro-negative metals iodine and fluorine, or to the electro-positive iron, calcium, sodium and potassium. The development of enzymology has identified the important metabolic role played by a number of other metals in minute traces in the body tissues, now known as trace elements. Modern technology has allowed their identification, analysis and measurement which have shown variations in individuals probably related to the foods, soils and geological strata on which they live. Cobalt, magnesium, manganese, copper, selenium, zinc and molybdenum are indispensable for biosyntheses and energy transformation. In certain areas of the globe, and within the different countries themselves, there are considerable variations of these substances in the soil and water. It has been suggested that these variations can lead to degenerative, neoplastic and other forms of disease: there is growing support for the contention that they can lead to regional patterns of disease, in the same manner that iodine deficiency leads to disturbance of thyroid function.

2. Specific Problems of Kuwait

To illustrate some of the thoughts which we have expressed we will review the problems of a small country which has undergone an extraordinary transformation in less than twenty-five years. The pace of change has however, eroded certain significant features of an ancient and proud people who are having to accept the responsibility of creating and maintaining a high speed commercial and industrial life in one of the hottest places on earth.

In Chapter II we have described the austereness of Kuwait's physical environment and it speaks much for the adaptability of those people from many parts of the world who have learned to control the unfriendly conditions by the development and use of the one major natural resource—petroleum—in less than twenty five years. This has not been without cost to the mental and physical health of these people and their families although counterbalanced by an almost unique approach to their welfare, first by the Kuwait Oil Company and followed quickly by an enlightened Government.

a) Training for responsibility

Until now the urbanized Kuwaiti of any substance has remained either in retail or import-export trade, but now many are rising fast in the higher echelons of the various Government departments or service industries, whereas the non-Kuwaitis, Arab and non-Arab, are providing the labour and technical services for which there is increasing demand. In Chapter V we reviewed the source of these immigrants, to some of whom the environment of Kuwait was no novelty, whether they came from the opposite shore of the Gulf or the oil-fields of Texas. But to many, Indians, Lebanese, Palestinians, and Europeans the new life presented many harsh aspects, which first had to be understood and respected if their stay in the country was to be reasonably comfortable and profitable.

b) Infectious disease

Because of the overwhelmingly dry nature of the climate with low relative and absolute humidity, the absence of surface water and drainage and the desiccation of potential breeding grounds for flies and other disease-carrying vectors, infectious disease is not the serious problem universally found in the humid tropics and inextricably bound-up with poverty (Fig. 39). While the absence of humidity and rain is a natural phenomenon, the avoidance of poverty has been tackled by the Government from the start of the petroleum era, and conditions in Kuwait, while comparatively unfriendly to man, present the same challenge to his traditional enemies, be they microbes, insects, reptiles or the more obvious mammalian predators. There is one important exception however: Kuwait from far back in its history has been over-ridden with rats, and many have been the occasions in the past when bubonic plague has ravaged the population. This is related to the sea communications between Kuwait and the great endemic centres, particularly India. Surprisingly, in the other well-known human diseases where the rat acts as the primary host and its parasites as intermediary hosts, it seems that the nature of the environment interferes effectively at some point along the chain of transmission, be they spirochaetal (relapsing fever), or rickettsial (typhus).

There may be one major point of significance here: until very recently, Kuwait City has not had a piped sewerage, which provides, of course, excellent pathways for a rat population to penetrate and proliferate. On the other hand although the town of Ahmadi has had such a system for twenty years the care and control of its use has been a relatively easy matter of public health engineering in view of the nature of the controlling authority and housing inspection. Recently the system has been connected with the new town of Sabahiya and the rapidly growing coastal area of Fahahil; the risk of rat migration now becomes a real threat. But the comparison of the

hostility and barrenness of Kuwait for the rat may be made with the older areas of Basra, where, even in daylight, the movement of rats along the steep banks of the city's canals is brazen. Basra has long experienced endemic rat-borne disease.

Kuwait has maintained until now relatively good environmental defences against infectious disease caused by the higher organisms, bacteria, spirochaetes and protozoa. This is exemplified by the comparatively low prevalence of typhoid fever and related waterborne bacterial diseases, non-venereal syphilis and relapsing fever, and clinically active intestinal amoebiasis respectively, in contrast with the virus diseases. Measles especially is a scourge and results in a high children's case fatality rate in certain groups; neither does influenza show any willingness to modify its sometimes severe effects as evidenced in the pandemic of 1957 [12, 13]. More recently the small-pox epidemic of 1967 provided a sharp reminder that the ambient conditions of Kuwait today are a ready stimulus to virus propagation and spread [14] which depend for their success on air and dust-born transmission, both freely available.

Nevertheless the comparative protection against non-viral agents is unstable and any change in the ecological balance can reverse the position. For example, whereas cutaneous Leishmaniasis, caused by the protozoon Leishmania tropica, is occasionally seen, visceral Leishmaniasis, induced by the closely related Leishmania donovani, has never, until recently, been reported as having been contracted within Kuwait [15]. The vector agent, a sandfly (Phlebotomus papatasii) has apparently established itself sufficiently securely to carry the organism from the natural or primary host, the dog, to the human. What has come about to facilitate this? The breeding of another sandfly (Phlebotomus sergenti), observed in Kuwait in rodent burrows in the desert and in moist areas near water taps, is the vector of cutaneous Leishmaniasis, but its habitats are either remote from dwellings or susceptible to insecticides if in the urban or shanty areas. But while P. papataisi is known to frequent broken down buildings where moisture and shade combine, that possibility was not considered before the redevelopment of large areas of the city, with the dumping of rubble just outside the city boundaries. This has apparently provided the right conditions. These rubble piles make excellent breeding grounds for the sandfly and are visited by the dog population for a variety of purposes. This exemplifies the need for planners and developers to organise their activities with the help of the epidemiologist, to avoid the danger of introducing serious new diseases.

The Arab has traditionally regarded the dog as unclean and not without reason: so much of the teaching of the Holy Koran and the Old Testament is based upon the simple hygiene experience of the early Semitic peoples, and while they do not always agree, for example on the value of wine, the pig and dog are frequently condemned. While the eating of pork is prescribed for all Muslims, the use of dogs has, until recently, been restricted to guarding encampments and for hunting, for which the decorous Saluki is the best known example. This dogmatic antipathy is changing now with the assumption of Western-pattern social status, particularly among the Arab immigrants and dog stealing has become a minor problem in Kuwait today. In any community, nomadic or settled agricultural, where sheep, goats, draught animals such as oxen and camels, and milch cows are cared for in proximity to humans and dogs, the intestinal parasitic tape-worm of the dog, Echinococcus granulosus, will infect cattle or man, who may ingest the eggs excreted in the stool of the dog. But instead of inhabiting the lumen of the small bowel of cattle and man, the larvae burrow through the wall and grow into large cysts in many different parts of the body, having been carried there by the blood stream. In Russia, Australia, New Zealand, South America, Spain, within the Arctic Circle and scattered throughout the world in smaller sheep-herding communities, the presence of these Hydatid cysts can present a diagnostic problem and will require expert treatment. Kuwait appears to have one of the highest incidences of new infection, at present 4%/o [16]. It is estimated that ten new cases are seen yearly [17]. Thus Kuwait has become one of the centres at which to study this interesting, disabling and often dangerous disease.

Prior to large-scale immigration into Kuwait, infections or infestations by the blood flukes, Schistosoma haematobium and S. mansoni, species of trematode worms, were rarely seen or at least seldom diagnosed. This was because any newcomers to Kuwait were either seamen or desert dwellers, only a small proportion of whom might have become infected, usually in childhood, at one of the bigger oases in Saudi Arabia or Iraq. The life-cycle of the parasite reveals the extra-ordinary persistence and adaptability of the human blood fluke in maintaining itself and seeking out its natural hosts, despite apparently impossible barriers. It does this by the spiny microscopic eggs which have been laid in the blood vessels of the gut, first making their way through the vessel walls and the tissues, to the cavities of the urinary bladder in the case of S. haematonium and large bowel in the case of S. mansoni. They are then passed externally with the urine and stool respectively and can only survive if these acts are performed into non-saline water, where the eggs hatch; a minute swimming form seeks out a particular tiny snail and develops further. Thus far they require water and a snail and these two are found through out the irrigated agricultural areas of the Middle East and elsewhere. After further development in the snail, the larval fish-like form then seeks out humans who are either bathing or working in the water, burrows through the skin into the tissues and blood-vessels to develop into the adult stage. By this stage then, we see that the requirements for continued propagation include human hosts urinating or defaceating into irrigation channels and the intermediate host snail. Neither of these were present in Kuwait in former days, but with the immigration of people from Southern Iraq and Iran, and more recently from Eastern Saudi Arabia and the Yemen, the human host is now quite frequently seen with symptomatic disease which requires treatment. One interesting source of these patients is the Deir-es-Zor region on the upper Tigris River just inside the Syrian border from Iraq, from which incidently, Dr. Hudson has so eloquently described the interesting condition called Bejel, a form of non-venereal syphilis [18]. With the rapid expansion of housing in Kuwait (Figs. 40 and 41) and the employment of young men, Kuwaiti and non-Kuwaiti, in a host of minor trades and indeed in the oil industry itself, a demand for wives has arisen which cannot be satisfied from Kuwait sources alone, especially because of the high cost to the prospective husband. The Deir-es-Zor region does not raise these pecuniary barriers and the girls are sought in marriage by the young men from Ku-

wait. No sooner do many of them arrive in Kuwait than they suffer an acute flare-up of their chronic schistosomal infection acquired in childhood, now no doubt related to the conjugal activities in which they quite understandingly indulge. However, the net effect is often the considerable annoyance due to discomfort in the uro-genital tract on the one hand, and the difficult eradication of the chronic disease by the medical authorities on the other, which may delay the arrival of children, the main reason for marriage. This is an instance of sociologically introduced disease in Kuwait.

At the present time neither the water for irrigation channels nor the snail are available in Kuwait for the further propagation of the disease, but the possibility has certainly been in the mind of the authorities when they have been considering future irrigation projects, in particular the Shatt-al-Arab pipe-line scheme. The tributary areas of this river in Iraq have one of the worst records in the world as an endemic region, due to the presence of the snail intermediate host.

The first clinical research unit to be established in Kuwait is devoted to disease of the alimentary tract. Situated at the Amiri Hospital it is directed by a physician who has had wide training both in the Middle East and the United Kingdom: so far studies [19—22] have been reported of some of the prevalent conditions affecting the colon and rectum and it is perhaps here that we can compare the effects of the more traditional diseases such as infection with Entamoeba histolytica (amoebic colitis and hepatitis), with the syndromes apparently related to the stresses of modern living, in particular the non-infective ulcerative condition characterised by a proctitis rather than colitis as usually seen in the West. Psychological disturbances, the excessive use of drugs and the eating of tinned food and delicacies are regarded as additional predisposing factors to the stress of the changing way of life in a previously simple people. This phenomenon is now being seen frequently in other countries on the march to development, but until now the Arabian peninsula has not had the non-infective condition recorded. Complaints of abdominal symptoms are frequent among these peoples and it is imperative to rule out the infective causes of bowel disease and avoid the injudicious over-prescribing of antibiotic drugs. The simple man or woman will frequently identify his troubles, often of a non-organic nature, by reference to the abdomen and we know that in more sophisticated countries we see the serious diseases such as duodenal ulcer, regional ileitis, and flamboyant ulcerative colitis for which we have not yet found a definite cause. In Kuwait there is indeed fertile ground on which to study the transition patterns and this is in fact being undertaken with enthusiasm, and exemplified by a recent investigation into poor function of the colon [22], in which some interesting aetiological factors were identified among 105 patients.

1. Psychological:
   Lack of security
   Family difficulties   } 74%/o
   Financial anxieties
   Fear of cancer

2. Infective colitis
   Amoebic
   Bacillary   } 44%/o
   Schistosomal

3. Food sensitivity   6%

4. Purgatives   6%

5. No obvious factors   10%

Salem [22] notes the interesting reversal of the predominance of symptoms among women which is found in Europe, and comments that this might be related to the more passive social role of women in Kuwait society whereas most responsibilities fall on the men. "It seems that psychological factors play a big role in the aetiology of the condition in both sexes of all social classes. In addition to the stresses of modern life, the climatic inconveniences add an extra burden on unacclimatised persons. Men of all nationalities are equally exposed to such strains but the condition is more frequent among Palestinians of both sexes: the hardship they have suffered might have a role ... it is postulated that the change of defaecation habit from the squatting Eastern position to the sitting Western one may play a role in developing redundant loop and pseudomegasigmoid that have been noticed in the higher income class of both sexes."

Throughout all countries without high levels of community hygiene, piped pure water supplies and widely varying standards of housing and health education, infectious diseases of the gastro-intestinal tract are an important cause of infant mortality, chronic childhood morbidity, malnutrition and underdevelopment. The wealth of any developing country depends upon the health and nutrition of its children and this is perhaps emphasised in Kuwait where the Government is very conscious of the inverse population ratio between native Kuwaitis and immigrants. Every encouragement is given to Kuwaiti families to increase their size.

c) Infant gastro-enteritis

To this end special thought was given by the paediatric services to infections of the gastro-intestinal tract at the beginning of the nineteen-sixties and parallel studies were initiated by the doctors in the State Medical Service and the Kuwait Oil Company. Startlingly dissimilar findings were reported particularly the fatality rates. These appeared to be related to the different populations and the lines of communication governing early treatment, rather than any differing factors once the child had reached hospital. Throughout the Middle East since the end of the 1948 Arab-Israeli war, the need for early rehydration has been recognised and pioneered by U.N.W.R.A. in Jordan and Lebanon, and by government sources in Iran in areas where hospital facilities are either not available or access to them is delayed. These "rehydration centres" have been successful in combating the terrible wastage resulting from this common complex of diseases.

In Kuwait there was agreement that the peak incidence occurred towards the end of the year, namely autumn, during the short fly-breeding season. All the investigators agreed that the importance of the nutritional state and complicating parasitic disease should be stressed as well as the failure to identify any outstanding specific factor, for in as many as fifty per cent of the infants, no pathogenic organisms were found in their stools. Nevertheless, whereas the most frequent organisms recovered in Kuwait were Proteus morgani and Salmonella newport,

in Ahmadi Escherichia coli was more than three times as common as Shigella and Salmonella. Again, the different social pattern and environment in Ahmadi may help to explain the discrepancy. Re-infection following recovery was not infrequent and reflected the importance of domiciliary follow-up and health education. One of the studies illustrated the well-known association between the location of dwelling houses and foci of infection originally identified by the predominance of patients who came from a living area through which the main Ahmadi town drainage passed. The sewers by nature of their construction could be damaged and become open to penetration and breeding by flies.

The social factors governing the risk of infantile gastroenteritis were those of housing, income, literacy, size of family, method of feeding, relation to nearest clinic or hospital and nationality. There was evidence that the delayed and inadequate weaning practiced by Kuwaiti mothers contributed to the increased incidence of diarrhoea in the Kuwaiti infants, particularly Badu, for in Shaker's series it led to both anaemia and hypoproteinaemia, fertile ground for gastro-enteritis.

Shaker Y. [23]

Total patients   940

(No. Kuwaitis   499 or 53.1%)

Ffrench G. [24]

Total patients   59

(No. Kuwaitis   19 or 32%)

The importance of early recognition, supportive and specific treatment were emphasised by the contrast in fatality rates between those cases arising in Fahahil town near Ahmadi who were eventually admitted to a hospital in Kuwait, thirty miles away, and those cases arising amongst oil company employees' dependants who were admitted to the Southwell Hospital in Ahmadi within a short time of onset.

Number of deaths Government series (Shaker, Y.)	Fatality rate %		Number of deaths Oil Company series (Ffrench, G.)
67	32.9%	1.6%	1

While the Government series came from an unselected population that of the oil company, by virtue of housing, income, controlled community and other factors could be regarded as partially selected and therefore statistical comparison would not be justified. Nevertheless, to take the example of Ahmadi, where at the time there was an estimated infant population of 2000, the calculated incidence of 30 per 1000 very severe cases per year is low. It was reckoned that for every case admitted there were ten treated in the early stages at the general practice clinics. The lesson learnt was that with education and health control in the homes, the incidence of serious disease could be reduced still further. There is certainly sufficient justification for encouraging early attendance at clinics even at the risk of over-loading them, for it is here that preventive paediatrics can be practised successfully. Such a disease as infantile diarrhoea can also be used as an indicator of the general level of hygiene and health in a developing community comparable to still-births; it should be studied along with population density, family size, and planning and location of housing. In many villages in Kuwait the continued practice of confining domestic animals, particularly sheep and goats, in the house and compound after sundown no doubt contributes to poor standards of hygiene and it will require considerable enthusiasm and persistance by the health authorities to discourage these traditional habits, remnant of a Badu past.

d) Kuwait as an epidemiological listening post

That Kuwait can be regarded as a listening post or monitor for conditions not previously recorded in the populations of the Middle East, is due to the enthusiasm and skill of its doctors and other health workers. Increasingly, reports are appearing which may help to elucidate the obscure causes of some of these illnesses, an example of which is sub-acute sclerosing pan-encephalitis, a form of brain tissue inflammation. Six patients were seen within a recent period of five years with this progressive and often fatal inclusion-body encephalitis suggesting a relatively high incidence in the population [25]. That it is possibly an environmental influence rather than genetic or social is suggested by its occurrence throughout such a mixed population of Arabs, in contra-distinction to another nervous disease called Kuru [26], recently studied among a primitive and isolated population in New Guinea.

This brings us to consider the patterns of ill-health influenced by the convergence of racial strains in a rapidly developing urban community.

e) Genetic disease

In any collection of animals, human or otherwise, in-breeding will eventually lead to disaster unless selection is practised at the same time. A recent attempt to do this in the human race was fortunately thwarted but not before a great and costly war had been fought, and therefore it is natural selection upon which humanity has to rely. Natural selection among palaeolithic and segregated peoples was provided by the principle of survival of the fittest but once the natural geographical barriers were removed man began to mix his genes and gradually the characteristics of a single primitive population were lost to merge with those of neighbouring peoples. The continuous influence of an unfriendly environment whether climatic or rendered unfriendly by microscopic or ultra-microscopic agents, required an adjustment of the body's internal constitution and defences to allow survival and reproduction. This came about by the selective action of gene mutation to enable the organism to withstand the insults to which it was subjected. Not all mutations provide this advantage, but only those which have given their possessor a unique malleability during the time he continues to be exposed. Nowhere is this better illustrated than by the fascinating pattern of haemoglobin molecules which are found today across the length and breadth of the world which is still endemic with malaria, or has been so within recent times.

Today there seems to be an increasing degree of interplay among peoples from East, West, South and North, bringing with them the genetic markers of their origin as well as the more obvious phenotypes, the recognisable ways in which the genes express themselves in the in-

dividual. The more recent appearance of recessive characters for instance may arise because of the better chance of survival of the homozygous state in the new community, whereas in the more primitive one, the social and medical conditions do not allow effective human survival or propagation. At the same time consanguineous marriage among the national Arab groups remains an important social cause of hereditary genetic abnormalities and great efforts in health education must be made to reduce this practice.

By the mid-nineteen sixties it was realised that unless study of these new problems was begun, much effort would have been wasted and little possibility of some future control could be expected. A unit of Medical Genetics was established in 1966 at the Sabah Hospital where both clinical and laboratory studies are undertaken and a careful record of family relationships is being built up, to enable gene frequencies to be estimated.

In co-operation with other departments of the medical services, emphasis has been placed upon the following priorities:

> The Haemoglobinopathies
> Mental disease
> Blindness
> Congenital malformations, of which perhaps the most important are:
> Congenital dislocation of the hip (G.D.H.)
> Congenital heart disease

This table of priorities lists the conditions which are now seen and which increasingly consume the time of the health authorities in treatment and social care, a phenomenon experienced among all socially developed communities. In the case of Kuwait this has arisen simultaneously with efforts to create a modern society out of a hotch-potch of peoples: this inevitably puts a great strain on the available resources. A study [27] from which the figures are taken was carried out during one year, 1967, in the maternity section of the Sabah Hospital and involved 4593 mothers who delivered 4625 babies, representing 23% of the total deliveries in the State of Kuwait during that period. Of these, 2220 were Kuwaiti mothers and 2372 non-Kuwaiti Arabs; most of the latter had arrived in Kuwait during the preceding ten years.

There were 32 pairs of twins (7/1000 deliveries) and 99 still-births (21.2/1000). The new-born infants showing malformations was 104 (22.5/1000) and 15 of these were still-born, an incidence of 153/1000 still-births. Males predominated (M/F — 108/100) with malformations slightly more frequent in the males (M/F — 112/100). Consanguineous marriages had a marginally higher incidence of malformations, 52.8% of the parents. The commonest malformations were neural tube defects (28 or 27%). The authors of the study point out that these figures were from hospital deliveries which usually give a higher incidence than the overall figure for the population. They also stress that other factors affect the significance, in particular the failure to recognise certain conditions until several months after birth, such as heart and kidney disease. Only "major" conditions were reported and confirmed the previously suspected frequency of consanguineous marriages and the youth of the mothers. The former is among the highest in the world for it excels the W.H.O. reported highest figures (Egypt) of 34%. These parents had a malformation rate of 26.2/1000 as against 19.5/1000, but in unrelated couples there was not such a distinctive difference in young mothers, 22.1% against 25.4% for the whole.

While the malformation rate in Kuwaitis and non-Kuwaitis was approximately the same, the total incidence (22.5/1000) was very close to those in Johannesburg (22.5/1000) and Panama (20.8/1000) [28], the two highest figures so far reported. The failure to find a significant difference between the ethnic groups, 21.6/1000 in Kuwaitis and 23.6/1000 in non-Kuwaitis, suggests that the environmental teratogenic factors played a greater role.

f) Environmental influences

The authors discuss the possible environmental influences, among which are the heavy consumption of supposedly therapeutic drugs in Kuwait related to the free medical services and the possible over-exposure to X-radiation: in 1967, out of a population just under 500,000, 350,000 radiographs were taken and several thousand fluoroscopies were performed. The effect of virus infection upon the developing foetus was considered, because of the small-pox epidemic and mass vaccination which took place in April—May 1967. They noted a seasonal incidence of neural tube malformations with probable peaks in June and January and suggest the presence of environmental factors operating on some foetuses conceived in the spring and autumn.

g) Blindness

The example of blindness can be taken to illustrate modern Kuwait's situation vis-à-vis the mixing of populations and changing disease patterns. "Blindness is the final disastrous result of ocular disease" [29] and nowhere can this be demonstrated better than in the Middle East. Nevertheless when comparing the situation with that of other Arab countries the relative effect of blindness in Kuwait is trifling. This does not detract however from the need to identify the causes, to treat and care for those already afflicted to prevent the possibility of a breakdown in the apparently stable situation. The evidence shows that the prevalence of blindness in Kuwait is on a par with that of the world generally, 0.5%: in more backward countries, less well endowed with environment and the means to prevent and combat early disease, this figure rises to 10% as in some parts of Africa. What contrasts from the pattern of disease in these countries is that in Kuwait, the number of blind persons below ten years of age constitutes a very small proportion whereas those above fifty, mostly indigenous Kuwaitis, are the majority. The rapid development of social and health services must have contributed, for particular emphasis has been laid upon the control of trachoma, purulent ophthalmia, measles and smallpox—the epidemic of 1932 has left its mark upon the eyes of the older age group—but the identification of glaucoma as an important cause emphasises that while disease patterns may change in a community the total weight of disease does not lessen; with increasingly longevity it may even expand. The establishment of glaucoma clinics has now been recommended, for this form of blindness can be prevented or retarded. The rising tide of accidents to the eye has contributed greatly to the problem and the pattern of industry with its multiple small units only helps to resist any efforts by the authorities to introduce and encourage the use of protective wear.

In the following chapters further examples will be given to illustrate the character of the struggle between adaptation and stress, some of it peculiar to Kuwait's geographical situation but steadily becoming linked with those problems encountered elsewhere in the established urban and newly urbanising communities both in the Middle East and the world generally.

VIII. Preventive Medicine in Kuwait

1. Preventive Medical Service

Kuwait, along with her neighbours, is the heir to some three hundred years of attempts, initially by the East India Company of England, to control the ravages of epidemic disease along the shores of the Persian Gulf. The history of these early struggles, which admittedly were primarily designed to allow the flow of trade to continue, is told in the next Chapter.

a) Establishment: urban planning: water supplies; killing of animals

With the coincidental establishment of the government medical services and crude oil production, measures to protect the health and welfare of the different sections of the population were put in hand. Today Kuwait can boast a efficient preventive medical service which has helped to achieve significant reductions in the accepted indices such as maternal and neonatal death rates, infectious disease, blindness secondary to trachoma, and tuberculosis; at the same time increases in the growth and weight curves of its children have been achieved.

Certain features of Kuwait's geography and urban planning have assisted in mollifying the effects of some of the more familiar diseases: The absence of flowing water in pipes, roadside drains or in underground sewers has reduced the morbidity of typhoid fever to manageable proportions. The ground radiation temperature of the sun, which can be as high as 180° F (82° C) in the summer, rapidly dessicates any organic material lying about, such as carcasses, faeces and garbage of all kinds, thus discouraging fly-breeding and fly-borne disease. Certainly this does happen in shaded areas, but not to the extent found in less arid countries. When the killing of sheep on a massive scale was carried out on the beaches, in villages and in the Badu camps adjacent to tents, when defaecation in the open was the natural habit and when garbage was indiscriminately thrown into streets, it was surprising that Kuwait for so long bore the reputation among its neighbours of being salubrious. The searing sun and low humidity provided the answer.

With the rapidly increasing population attending the boom in oil production, particularly after the Abadan incident of 1951, since when output has increased by 100,000 tons every six years to reach more than half a

million tons today, the natural means of disinfection can
no longer be relied upon. Within the past three years a
sewerage system with marine and activated sludge disposal has been started together with piped drinking
water, both of them potential destroyers of local geographical immunity to disease.

The vigilance of the Public Health Authorities has been
commendable; the early legislation to control the killing
of animals and selling and handling of food contributed
to morbidity figures small enough to be handled by the
developing diagnostic and treatment services. The establishment of a Public Health Inspectorate, medical
examination of known food handlers, the vigorous use of
insecticides and the control of refuse and sewage disposal
have made Kuwait the envy of all well trained Public
Health workers in the Middle East.

b) Structure of Ministry of Public Health

The Ministry of Public Health, presided over by a
Cabinet Minister who is not a member of the Al-Sabah
family, has the following administrative structure (see
Table 32).

c) Growth of services

The accompanying table (Table 33) gives an indication of the growth in services and staff which preceded
the present comprehensive establishment. It should be
borne in mind that this growth, along with demands in

Table 33. *Growth of services*

	1949	1953	1957	1962	1966	1970
1. Veterinary Officers	—	1	5	11	13	
2. Veterinary Assistants	—	4	16	20		
3. Veterinary Establishments	—	1	3	5		
4. School Doctors (Special)	—	—	4	7	38	
5. School Clinics	23	41	74	128	178	
6. Preventive Health Doctors					11	
7. Preventive Health Centres	—	1	3	7	11	
8. Public Health Inspectors			4	21	30	
9. Health Educators						
10. Quarantine Doctors					8	
11. Quarantine Inspectors	—	13	35	42	84	

other Middle East countries, placed a severe strain on the
capacity of the new training schools; as a result recourse
had to be made to India, Pakistan and Europe.

d) Communications

The importance of good communications was not for-
gotten; in addition to a rapidly developing orthodox
telephone system, radio communications have been
developed linking the Ministry, hospitals, outlying clinics
and transport services. These, together with a dependable
road system, have reduced the difficulties of public health
control and transportation of patients to hospital. Perhaps the problems of the Haj pilgrimage and the efficient

Table 32. *Structure of Ministry of Public Health*

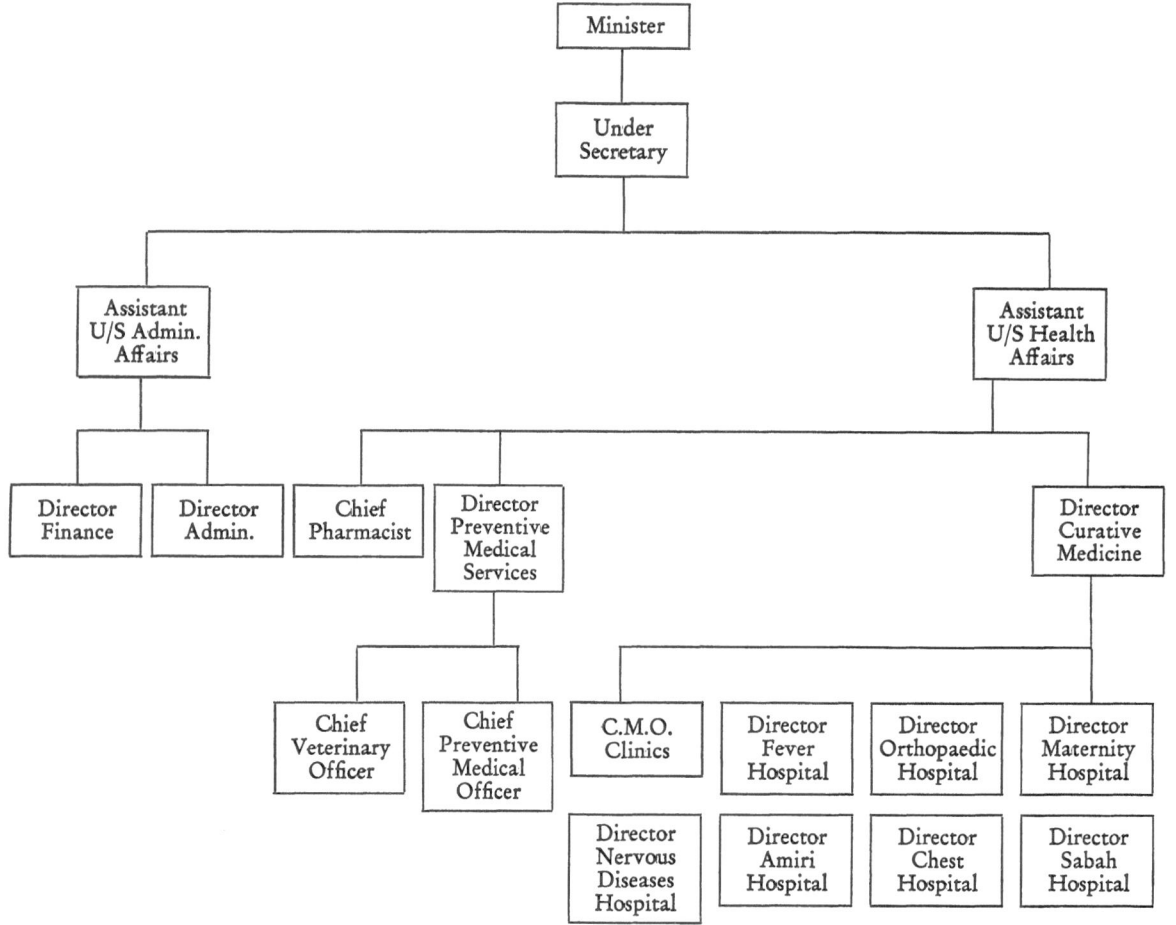

methods to tackle them exemplify the awareness by the Government of their responsibility to permanent and temporary residents within the borders of Kuwait.

e) Health education

Among the priorities in preventive medicine have been the dissemination of accurate knowledge of disease prevention, social welfare, marriage guidance and family planning, in this context the establishment of large healthy families. The use of the mass media, posters, magazines, pamphlets, newspapers, radio and television, have all been enlisted with outstanding success to the point when the clinic attendances, though vast, do allow the health authorities to reach the people, treat conditions early and give prophylactic advice.

The support given to the perennial themes of World Health Day from all sections of the community, Government, industry, trade and other organisations has been impressive. School teaching has tackled the problems of health and sex education to the point where youth is now pressing parents to provide it with conditions which they have learnt will secure them a disease-free existence, the result of an ideal policy which runs the risk of creating widespread anxiety and frustration; there is now some evidence of this.

f) Inspection of buildings and regulations

Along with health education have arisen new and hygienic shopping centres and bazaar areas (suqs), first-class modern abattoirs, food markets, bottling and food processing plants, all of which come under inspection. These, together with modern housing provided by re-development and the establishment of new urban areas in virgin desert, have contributed to the population expansion by natural accretion of Kuwaitis and others. The traditional diseases on the other hand have not yet shown the decline to be expected; the influence of better reporting and analysis may be the answer to this paradox (Table 34).

Table 34. *Improvement of prevalence of some infectious diseases*

Disease	1958		1962		1966	
	No. cases	Rate /1000	No. cases	Rate /1000	No. cases	Rate /1000
Typhoid	135	0.67	140	0.4	452	1.0
Pulmonary Tuberculosis					1200	2.5
Poliomyelitis	19	0.095	39	0.1	70	0.15
Meningococcal Meningitis	22	0.1	17	0.05	44	0.09

g) Private industrially financed medical services of the oil companies

At this point we interject a discussion on the value of private industrial medical services alongside Government service, their influence on the local community and the stimulus they may give to the country-wide medical services by example in planning, manning and organisation. The importance of full understanding between Government and company medical staff cannot be over emphasised; the mutual confidence which has developed over the years has contributed greatly to the stability and development of the total medical services.

Since 1960, when the first report on the Southwell Hospital of the Kuwait Oil Company was published [1] (Fig. 46), many changes have taken place in the State of Kuwait and in the oil town of Ahmadi, accompanied by a population expansion from 300,000 to over 500,000 in the former and from 15,000 to 25,000 within the Company's responsibility in Ahmadi. With this growth there has been a steady shift in the national composition of employees from non-Arab countries (British, Americans, Indians, Pakistanis and Iranians) to a predominantly Arab work force recruited in Kuwait and from the neighbouring countries of Lebanon, Jordan, Iraq and Oman. This changing pattern has brought with it a larger population of immediate dependants, wives and children, and the emphasis has now shifted to Arab family health care rather than health control of the industrial workers and expatriate dependents as formerly. The yearly admissions now total between five and six thousand and outpatient general practice and specialist attendances between two hundred and fifty and three hundred thousand. Gradually more accent has been laid on the preventive aspects of the work until it is apparent that prophylactic vaccination and inoculation, maternity and child welfare clinics, health visiting and education, tuberculosis control and insecticides campaign play a vital role. Together with this expansion has developed an occupational health service to deal with those problems peculiar to oil production and refining industries and their necessary servicing organisations. Such an approach has paid dividends because in spite of the larger population, for the past few years inpatient and outpatient attendances have been kept at a steady or even reduced level. While the local Ahmadi population has increased by 75%, the neighbouring coastal areas of Fahahil, Shuaiba and Fantas have exploded and from not more than 5,000 in 1960, now number more than 40,000. This has been occasioned by the Kuwait Government's choice of Shuaiba as the site for its industrial development plans, and the dormitory area of Subahiya to serve it.

Southwell Hospital has become the accident centre for the developing southern part of Kuwait; this has emphasised the need for keeping the facilities under review, particularly the introduction of intensive care and disaster planning. The need for both has been illustrated by more than one incident.

The Ahmadi population pyramid reveals a wide bulge in the very young (0—10) and the 30—50 age groups which at once shows where the weight of family general practice is felt. Maternity services are handling between 1,000 and 1,100 births in hospital, another effect of the urbanization of the previously scattered Company population who formerly lived in tents and wooden shacks. This concentration of responsibility allowed the closing of outlying clinics and the placing of maternity and dental services under one roof. Much better communication developed within the Medical Department, for instance between family clinics and records of individual immunity. Today the transport situation is no problem due to the high ratio of cars per population in Kuwait and particularly in the Ahmadi industrial area where it is one car to four people. The only complication arising from this is parking space, preferably out of the sun!

Thus the Southwell Hospital has ceased to be purely a hospital, with all the traditional implications that such an institution has in any community, but has become a Health Centre, providing health information on the prevention of illness, maternity services, family clinics, special follow-up and specialist clinics, an accident service, and finally a district hospital service equipped to handle medical, surgical, paediatric and isolation cases. The general construction of the hospital has lent itself, from the first, to modification into small and larger units, and much credit lies with the original planners.

An inestimable advantage has progressively accrued. It has become possible to integrate every member of the general practice Medical Staff into either inpatient or special clinic work, or both, which has added to their interest, keenness and therefore value to the community. The staff consists of physician, surgeon, obstetrician, gynaecologist, Medical Officer of Health, anaesthetist, ophthalmologist, radiologist, occupational health physician and thirteen general duty medical officers who also undertake the following practice, having obtained the necessary diplomas required under Kuwait State laws:

Dermatology	Obstetrics
Ear, Nose and Throat	Paediatrics
Anaesthesia	

There are also three junior medical officers for twenty-four hour emergency duties, which makes up the total professional staff of 22 doctors and 4 dental surgeons.

Such a system of medical duties has enabled flexibility and sharing of interests and would appear to be one answer to the problem of how general practice can be developed and integrated with a hospital service. Unless a doctor is stimulated and challenged in the wider fields of medicine, continual clinic and domiciliary work will cease to have an appeal. The above duties are carried out with the assistance of medical auxiliaries, whether qualified male and female nurses, or by nursing assistants who are trained in the department.

A continuous weekly programme of in-service training for all members of the medical and nursing staff is followed. All staff are encouraged to take an interest in the work of the Preventive Medicine Division and the occupational health service, not least for the good reason that the Medical Department exists primarily to support the industry of one of the world's great oil fields.

It is worth considering the value of an independent industrial medical service within a developing country. At the best it can give leadership, example and practical assistance to be followed by gradual integration and final withdrawal of independence and expatriate staff. It can train nurses and medical auxiliaries, initiate and take an active part in postgraduate medical education and research, and provide very practical evidence that industrial "exploitation" can be a partnership built on mutual respect and practical necessity.

2. Infectious Diseases

a) Comparison with other geographic areas

The national profile of infectious disease might be expected to vary with location, climate, ease of land, sea and air communications, social development and the quality of preventive medical care coupled with the degree of health education of the people. Kuwait, by virtue of its geographic position enjoys that hot dry or desert climate which is the antithesis of the equatorial rain-forest climate, and it is therefore of interest to compare the patterns of disease in Kuwait, technically outside the true tropics being north of the Tropic of Cancer ($23\frac{1}{2}°$ N.), and Uganda actually astride the Equator (Table 35) and find that in fact smallpox and poliomyelitis, the differences are probably not significant.

Table 35. *Infectious diseases.*
Comparison with other geographic areas

Geographical area	Kuwait	London Borough [a] of Ealing	Uganda [b]
Year	1967	1967	1967
Population	500,000	300,000	5,000,000
Disease Rate/1,000:—			
Chicken-pox	4	common	common
Diphtheria	0.1	nil	0.01
Dysentery	0.9	1.3	
Rubella (German Measles)	0.07		
Hepatitis	1	0.14	0.5
Leprosy	0.01	nil	Up to 0.01
Malaria	0.15	0.003	Up to 0.15
Measles	4	11	common
Meningococcal Meningitis	0.04	0.015	0.01
Mumps	4		common
Poliomyelitis	0.18	nil	very high
Scarlet Fever	0.05	0.41	
Small-pox	Epidemic 0.01	nil	Endemic up to 0.6
Tetanus	nil	nil	0.1
Trachoma			3
Typhoid	0.5	0.015	0.07
Whooping Cough	0.4	0.44	common

[a] Annual Report of Medical Officer of Health, Ealing 1967.
[b] Hall, S. A., Langland, B. W.: Uganda Atlas of Disease Distribution. Kampala 1968, p. 187.

The four main scourges are seen to be mumps, measles, chicken-pox and hepatitis; we will be discussing tuberculosis in a later Chapter. No special significance attaches to mumps, although the impression is that more adults are infected than is the case in Europe. Measles continues to be a killer, attacking young infants, the complications —principally pneumonia—being the main cause of death. Chicken-pox presents no problems which are not common experience, except during the 1967 small-pox epidemic, when differential diagnosis was vital and often difficult.

It will be seen that no tetanus cases were reported in 1967 [1]; in fact, the reporting of this disease has not been good, for when an investigation into the incidence was conducted in 1965 [2], 123 patients were found to have been infected between 1960 and 1965, although the annual official returns reported only half this figure— 68 cases. But the incidence is now rapidly falling, for most of the patients in the recent past have been new-born Badu; the absence of hygienic midwifery in the desert tent or sarifa coupled with the traditional practice of sealing the cut umbilical cord with camel dung accounts for the extreme youth of the patients. Elsewhere in Arabia the incidence is still high, where patients are brought to hospital from scattered oases such as in the Eastern Province of Saudi Arabia in the Arabian-American Oil Company's hospitals.

A mass campaign of immunisation against tetanus along with diphtheria, pertussis (whooping-cough) and poliomyelitis has been under way in Kuwait for some years, but the scattered Badu families are often out of reach of the government teams. More recently measles vaccine has become available and, although the susceptible population is still very sparsely covered, should begin to cut back the serious sequelae and mortality previously experienced.

An Infectious Diseases Hospital was opened in 1957 to cope with a rapidly rising tide of small-pox cases which had reached epidemic proportions. After the outbreak was over it was realised that a great need existed for the isolation and treatment of infectious disease; for over ten years this has been the State policy. A new Infectious Diseases Hospital of over 80 beds was soon required and by the time another small-pox epidemic hit Kuwait in 1967 the authorities were able to provide very good facilities indeed. The reports of the latest outbreak [3, 4, 5] are in many ways unique, both for the quality of the reporting and their illustrations which are outstanding; and they are the first record of small-pox in Kuwait which has been published, although by no means the first outbreak.

b) Leprosy

Leprosy is a small though continuing problem in Kuwait principally because of the difficulty in educating physicians to keep the possibility in mind. As in other non-endemic countries, the scarcity of clinical material dulls the alertness of the physician, despite the fact that many doctors in their own country have been familiar with the disease at some time in their training (Fig. 47). At present there are perhaps 20 patients undergoing specific treatment either in the Government Leprosy Hospital of 21 beds, or as outpatients. Tuberculoid and dimorphous leprosy predominate, as would be expected where all cases are imported, and present no particular problems once diagnosed, except for the handling of the interesting and unusual, though well-known, association of leprosy with tuberculosis in the same patient.

In a country with primitive pastoral peasants, such as the Badu, who husband their flocks of sheep, goats and camels, and who, with the onset of urbanization, have not yet discarded their habits of living closely with them, it would be expected that humans and animals might share the same diseases. This is true for Kuwait only in a very restricted sense, for the incidence of the zoonoses is extremely low. The occasional patient with cutaneous anthrax is seen; cutaneous leishmaniasis and brucellosis are rarely reported, and the arthropod-borne rickettsial and viral diseases have not been identified in Kuwait. A single patient developed visceral leishmaniasis while in Kuwait [6].

c) Hepatitis

Hepatitis being more frequently of virus origin rather than protozoal (*E. histolytica*) or toxic, presents no greater problem than that experienced in any country with good treatment services. The relative influence of infectious hepatitis virus and that of homo-logous serum hepatitis following parenteral infections and blood transfusions is not easy to assess but impressions are that at present the latter is not a serious problem. When sophisticated treatments involving kidney and heart machines become the rule rather than a rarity—only kidney machines exist in Kuwait at present for acute emergency treatment—then the possibility of serum hepatitis will be real, with danger to patients, medical and nursing staff alike. At present many thousands of visitors come to Kuwait without having taken prophylactic inoculations of human gamma-globulin and there is no increased risk of developing hepatitis than, for instance, in Europe. This is in distinct contrast to the situation in India. Amoebic hepatitis is not infrequently seen, mostly among returning expatriates from India and Pakistan. Diagnosis rests very much upon the appreciation of risk.

d) Rabies

Rabies has not been reported in Kuwait, either in animals or man. Strict requirements are in force for dog immunisation and dog control, with periodic destruction of stray dogs. Neighbouring countries, particularly Iraq and Iran, have endemic rabies and the danger of exotic disease must be guarded against. Increasing acceptance of the dog as a domestic pet, which has followed the immigration of peoples from western-orientated countries, has displaced the traditional recognition of the animal as unclean; previously dogs were kept only by the Badu for guarding camps. Every dog which behaves in an unusual manner should be or quarantined; this public health order is perhaps observed more in word than in deed and the danger of a sudden outbreak is always to be feared. Facilities for histological diagnosis of animal brains do not exist in Kuwait and material is sent to Cairo, Teheran or England.

e) Malaria

Malaria is an imported disease, occurring only in those coming from an endemic area, whether nearby in Iraq, Iran, Saudi Arabia and Oman or more remotely in the Yemen, Pakistan and India. It presents no problems other than the diagnosis, for most patients have a relapse of P. vivax or recrudescence of P. falciparum. Nevertheless the occasional acute hyperpyrexia in a small child may present and if malaria is not considered tragedy can ensue: This problem is not different from that encountered in many countries where modern travel has thrown down the barriers against imported disease.

f) Poliomyelitis

Poliomyelitis until recently has been highly endemic in Kuwait and the disablement arising from it can be seen still in the orthopaedic departments of hospitals and in the rehabilitation centres. New infections are still arising, chiefly among remote Badu families or immigrants who have not been immunised; while the actual number of patients with acute disease has remained fairly constant over the past few years, the incidence has not shown the expected decline; this probably once again reflects the improved reporting of new cases rather than any failure of the state-wide oral immunisation programme. Access to clinics and hospitals, earlier diagnosis, and modern treatment have combined to effect a significant reduction in the florid patterns of disablement so familiar in the days before the oral Sabin vaccine. Organised studies of virus isolations have not been made or rather not been reported, neither have conversion rates following the oral vaccine; only the future will tell us if those immunised will retain immunity, boosted by repeated vaccination or rendered ineffective by failure to take government advice. Absence from the country, particularly among those children who attend school outside Kuwait in Jordan and Lebanon and return to Kuwait intermittently, is another risk because of fallout from the programme.

g) Bacillary Dysentery

Bacillary Dysentery continues to present treatment problems, both in infants and adults. The prevalence of resistant organisms, mostly of Shigella type, is not known but is well recognised. The incidence of fresh laboratory-confirmed infections on the other hand is, when accurately reported, some 10 per 1,000 per year and accounts for considerable mortality among children of less well-endowed families [7]. Fortunately, for a number of reasons which are not clear, the post-dysenteric syndromes of arthritis and chronic colitis and the specific amoebic colitis are not common.

h) Amoebic Dysentery

Active colitis has to be distinguished from the cyst carrier, high rates of which may exist in Kuwait although no exact figures are available. The policy of treating all cyst carriers has been followed in the Kuwait Oil Company for many years on the premise that at any time the balanced immunity state may break down to produce overt active disease. In addition the cyst carrier is a public health hazard particularly if he or she is a food handler. The introduction of modern drugs such as Furamide (entamide furoate) and Flagyl (metronidazole) have made the treatment a simple outpatient procedure; it is advisable however, to treat in hospital to guarantee that the drug is taken and that resistance due to inadequate dosage does not occur. Amoebic colitis, when it occurs in Kuwait, is on the whole a relatively mild condition, responding well to either traditional treatment with emetine and the iodoquinolines, or the newer drugs, dihydroemetine [8], furamide and flagyl.

i) Worms

Perhaps the most important of all infectious disease, because of the frequent mild disability, is caused by worms: the modern term 'geohelminths' has been coined to emphasise the relation of the hosts, primary and secondary, the worm and the external environment, change in any one of which will create more or less unfavourable conditions for the survival of the worm. The capacity of the worm to adapt, remain in stable relationship with hosts and survive in unfriendly environments is one of the wonders of nature. Nevertheless the desert of Kuwait has not been kind to worms particularly those which rely upon an external environmental stage for development, whether as a free-living larva or as an egg. Thus the range of worm infestations in Kuwait is limited to those which do not require a potentially unfriendly transmission stage either from individual to individual or from the carrier to food and then to the new host.

Of the helminths and flukes found in humans in Kuwait the following table gives a general picture:

Table 36. *Helminth infestation in the population of Kuwait*

The parasite	Mode of transmission	Relative proportion
Roundworm (Ascaris lumbricoides)	Eggs to faeces to fingers to food.	100
Pinworm (Enterobius vermicularis)	Eggs to faeces to fingers to mouth. House dust.	20
Whipworm (Trichuris trichuris)	Eggs to faeces to mouth.	10
Hookworm (Necator americanus, Ankylostoma duodenale)	Larvae to faeces to ground, through skin to gut.	7
Strongyloides stercoralis	1. Larvae to faeces through skin to gut. 2. Larvae to faeces to ground, through skin.	5
Flukes (Schistosoma mansoni S. haematobium)	Eggs to faeces or urine to snail to new host.	10
Tapeworm (Taenia saginata)	Larval cysts in meat to human.	5

It will be seen that the roundworm *Ascaris* is the predominant intestinal worm of humans in Kuwait. For how long this has been so is not known, but there is a strong suggestion from elsewhere [9] that with increasing urbanization, the prevalence of roundworm infestation has increased despite the social, nutritional and educational advantages. There is no reason to doubt that this applies also to Kuwait, and it has been perpetuated by the importation of fresh vegetables since 1950 from the growing areas in Jordan, Lebanon and Iraq, where there is very heavy infection and human excreta has been used as manure; undoubtedly this provided a powerful impetus to the natural increase due to urbanization alone.

Hookworm is probably endemic in some of the oasis villages in eastern Saudi Arabia but this has not been proved in Kuwait where infection is never heavy; the eggs are usually discovered incidentally. By far the majority of patients would appear to have developed the infestation outside Kuwait.

Hydatid disease due to the encystment of the larva of *Echinococcus granulosus* is found occasionally and has been mentioned in the previous Chapter; because of the paucity of the cases, the diagnosis and, even more the treatment, requires considerable clinical acumen and expertise.

j) Schistosomiasis

Infection with the blood flukes *Schistosoma mansoni* and *S. haematobium* has also been given as an example of geographical pathology in the preceding Chapter. No figures of prevalence or incidence exist, but patients with the infection are always to be found being treated in any of the large general hospitals in Kuwait. All are examples of exotically acquired disease; Forsyth [10] reported a series of 217 patients seen in one hospital over a period of seven years spanning the first large immigration into Kuwait. Badu Kuwaitis have been found with the disease but the Badu does not restrict himself within national State boundaries and no foci of infection have been found in Kuwait.

The sequelae of the infections, which are due to the inflammation caused by migration of the eggs from the abdominal veins to the intestine *(S. mansoni)* and urinary tract *(S. haematobium)*, justify diagnostic screening and effective treatment. Forsyth [11] estimated the incidence of urinary bilharziasis (schistosomiasis) to be roughly one in 15,000, listing calcification of bladder and kidney stones as a frequent accompaniment. He also observed the more complete destruction of ureter and kidneys seen in the highly endemic areas. Treatment is, however, usually successful in preventing viable eggs doing further damage. Several authors [10, 11, 12] have agreed that the absence of re-infection within Kuwait creates a unique opportunity to observe the value of different methods of treatment. At the same time, the facility of being able to treat all patients in hospital has a considerable advantage over the more casual outpatient regimen in hyperendemic countries. Cure rates of 99 per cent have been reported [10]. Opportunities for major reparative urological surgery will presumably continue to exist for a number of years.

The moral to be drawn from this description of infectious disease in Kuwait and indeed from elsewhere in rapidly developing and urbanizing communities, is that health education, sanitary engineering and avoidance of overcrowding together with early diagnosis are the essential ingredients for containment, control and finally elimination of at least some of the dangerous forms of infectious disease.

3. School Medical Services

a) School population

In the past thirty years the student population of Kuwait's schools has increased by a factor of more than fifty, from 2,000 in 1940 to nearly 120,000 in 1970 while expenditure has grown from less than 40,000 K. D. to more than 25 million K. D. annually, a factor greater than 600 (Fig. 48, 49, 50). The health of today's school child and adolescent is incomparably better than a generation ago; much of this can be claimed by the general social and medical services but undoubtedly it was the foresight of the early administrators following the first production of oil twenty-five years ago in establishing school medical services, which has done so much to identify and control infectious and nutritional disease. Prior to 1950 there were no school doctors or clinics; today the comparable figures are 38 and 178 (1966), to serve the kindergarten, primary, intermediate and secondary stages of the student's life.

b) Development of service

The first school doctor was appointed and a medical service commenced in September 1950 and in the following month a woman doctor was recruited for the girl's schools. Very shortly afterwards, with the active cooperation of the newly appointed ophthalmologist, anti-trachoma measures were instituted in the schools, measures which have paid a handsome dividend, for with the reduction in partial or complete blindness and the progressive extension of educational facilities, the literacy

rate is now high among people below the age of forty. This early experience of the school health service in identifying and treating infectious eye disease determined the future pattern of its work and it has continued to give both preventive and curative services.

Every school, of which there are now over two hundred, has its own clinical unit with full-time nurse and visiting doctor three times a week. Selim [1] has estimated that every 1,000 students receive about 155 hours of medical care at the school during the academic year. Each student has a health card which he retains during his school career: On this are recorded the results of periodic physical examinations, inoculations and vaccinations together with details of investigations and treatment. These records have enabled the permanent team of school medical specialists, not only to consult on individual health problems, but to plan and carry out a number of important surveys, among them ringworm infection of the scalp, trachoma, tuberculosis, dental caries, skin and venereal diseases. The latter two [2] are worth reporting in detail for they give a unique picture of a continuing epidemiological problem among youth confronting both settled and developing countries.

c) Epidemiology

In Kuwait the students under survey came from the following ethnic and national backgrounds:—Kuwaiti, Iraqi, Syrian, Jordanian, Palestinians, Saudi Arabia, North African Arabs, Lebanese, Yemen and South Arabian Arabs, Arabian Gulf Arabs, Iranis, Indians and Pakistanis. They attended the skin and venereal disease outpatient clinic of the school health service over a period of 8 years, the total visits being 85,000 among 15,450 patient records (9,789 students and 5,661 teachers and other school employees).

These are imposing lists (Table 37—39) and illustrate the very varied superficial infections, and scaly and otherwise hygienically unpleasant conditions which may abound every day in our schools. The responsibility of educational and medical school authorities is clear in the need to identify, segregate and cure. Text Fig. 22 confirms once again the reduced incidence of many diseases in Kuwait during the hot dry weather; an example is given of oriental sore of which the authors saw 53 cases with a seasonal incidence from November till June, maximal in January and February. The close coincidental association of papular urticarial lesions with oriental sore suggests the possibility of common vectors, but no field studies have yet been made.

The observation of alopecia areata, lichen planus and leucodermia occurring in the same patient and sometimes in more than one member of a family is of speculative interest. The absence of other diseases such as leprosy, lupus vulgaris, sarcoidosis, tuberculoid eruption and skin manifestations of metabolic disorders was notable.

When we turn to the venereal diseases we see that there were 212 cases of gonorrhoea in males only, an incidence of 1.38% of the total population studied. While the patients are not specifically identified as students, that a large proportion of them are, is implied. This together with the diagnosis of 15 student cases of early syphilis (Table 39) in the final three years of the study has occasioned considerable concern but no doubt reflects the licence now taken by the youth of a highly developed socially mature state and perhaps we can in-

Table 37. *Incidence of various skin diseases*

Diseases	No. of females	No. of males	Total No.	% in females	% in males	Total %
Superficial fungus infections of skin	479	988	1467	3.07	6.34	9.41
Allergic skin diseases	1181	1151	2732	7.65	10.09	17.74
Erythema multiforme	38	54	92	0.25	0.35	0.60
Erythema nodosum	9	4	13	0.06	0.03	0.09
Pityriasis rosea	133	127	260	0.86	0.82	1.68
Psoriasis	30	54	84	0.19	0.35	0.54
Discoid lupus erythematosus	2	7	9	0.01	0.05	0.06
Solar dermatitis	11	22	33	0.08	0.14	0.22
Prickly heat	111	156	267	0.73	1.01	1.74
Impetigo	1130	1473	2603	7.32	9.54	16.86
Seborrhoeic dermatitis	97	133	230	0.64	0.86	1.50
Acne vulgaris	858	762	1620	5.56	4.94	10.50
Pityriasis capitis	590	329	919	3.83	2.14	5.97
Diffuse hair loss	501	194	695	3.25	1.25	4.50
Trichomycosis axillaris	2	3	5	0.01	0.02	0.03
Oriental sore	25	28	53	0.16	0.18	0.34
Other parasitic diseases	57	78	135	0.37	0.58	0.88
Warts	545	1070	1615	3.54	6.67	10.21
Molluscum contagiosum	22	60	82	0.14	0.38	0.52
Herpes simplex	75	113	188	0.48	0.72	1.20
Chicken pox	180	243	423	1.18	1.57	2.75
Herpes zoster	26	60	86	0.17	0.38	0.55
Measles	5	5	10	0.03	0.03	0.06
German measles	10	13	23	0.07	0.08	0.15
Scarlet	25	25	50	0.16	0.16	0.32
Chilblains	53	24	77	0.34	0.15	0.49
Sclerodermia	2	2	4	0.01	0.01	0.02
Alopecia areata	84	142	226	0.55	0.92	1.47
Lichen planus	34	57	91	0.22	0.37	0.59
Leukodermia	114	96	210	0.74	0.62	1.36
Prurigo nodularis	9	18	27	0.06	0.12	0.18
Dyshidrosis	64	83	147	0.41	0.53	0.94
Hyperhidrosis	15	22	37	0.04	0.14	0.23
Chloasma	60	5	65	0.39	0.03	0.42
Parapsoriasis	0	4	4	0.00	0.03	0.03
Pityriasis rubra pilaris	5	6	11	0.03	0.04	0.07
Follicular hyperkeratosis	62	64	126	0.4	0.41	0.81
Hypertrophic skin manifestations	136	153	289	0.87	0.99	1.86
Keratolysis exfoliativa	63	110	173	0.37	0.71	1.08
Nail affections	39	21	60	0.25	0.13	0.38
Tylosis	14	9	23	0.09	0.06	0.15
Icthyosis	0	18	18	0.00	0.12	0.12
Gonorrhoea	0	212	212	0.00	1.38	1.38

Table 38. *Incidence of the various allergic skin disorders*

Diseases	No. of female Pts.	No. of male Pts.	Total No.	% in females	% in males	Total %
Urticaria	221	264	485	1.44	1.71	3.15
Papular urticaria	248	428	676	1.6	2.83	4.43
Drug rash	16	49	65	0.1	0.31	0.41
Angioneurotic edema	21	37	58	0.13	0.24	0.37
Atopic dermatitis	67	51	118	0.43	0.33	0.76
Contact dermatitis	325	292	617	2.11	1.9	4.10
Eczema	285	428	713	1.84	2.77	4.61
	1181	1551	2732	7.65	10.09	17.74

Table 39. *Incidence of syphilitic infection*

	Student Pt. Kuwaiti	Non Kuwaiti	Non student Pt. Kuwaiti	Non Kuwaiti	Total
Primary Syphilis	6	–	1	6	
Secondary Syphilis	8	1	2	5	
Total number	14	1	3	11	29

dicate this trend when the relative preponderance in Kuwaiti students is contrasted with the incidence among the non-Kuwaiti students. There has been no report of the diagnosis of non-venereal syphilis (bejel). The possible association of venereal disease with alcoholism in young Kuwaitis should be considered; this problem was one of the governing factors in deciding the Government to introduce prohibition of all alcoholic drinks in 1964. Unfortunately, as experience elsewhere has shown, prohibition has merely acted as a spur to illicit production and sale together with large imports by smuggling.

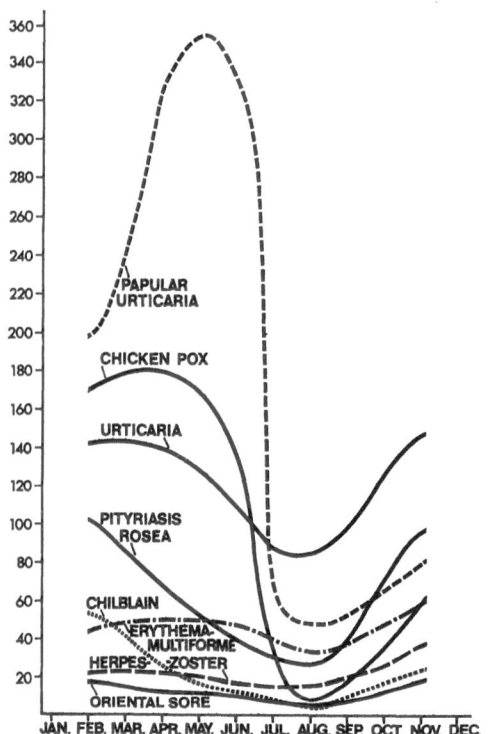

Text Fig. 22. Seasonal incidence of certain diseases

But it would be unfair to leave our subject on this unhappy note for the sins of the few do not match the achievements of the many. The health education programmes initiated by each school have combined to reduce to manageable proportions the diseases and disorders of children and promote a healthy and enquiring mind in a healthy body.

d) Nutrition

To this end an outstanding contribution has been made by the attention to school diet. A central city school kitchen prepares food, at least one meal daily, for over three-quarters of the student population and has proved a major factor in the upgrading of nutritional standards. Such a centralised food supply as this must always have a sword of Damocles over its head, for any deviation from strict hygienic methods may distribute infectious disease rapidly to a very wide and susceptible population. This potential was unfortunately realised in 1967 when an outbreak of staphylococcal toxin food poisoning spread like fire through several schools creating great alarm among the townspeople, and severe effects in many children. It was traced to a food-handler in the central kitchen with an infected hand.

IX. Treatment Services

1. Traditional Medicine

Traditional Badu medicine in Kuwait has not had very much attention paid to it, but from experience so far gathered it does not appear to play a significant part in the present changing culture. Dickson [1] has described the dozen or so indigenous practices of which he had experience some years ago but it is doubtful if any of these are pursued with any enthusiasm today, now that the benefits of Western medicine are known the length and breadth of this small country. Yet Zahra Freeth [2], Dickson's daughter, remarked on the fatalism of the Badu towards sickness and death and the reluctance to seek modern medical help only a bare twenty years ago. This contrasts with current demands which create such a heavy strain upon the existing services.

a) Badu practices

The following prescriptions are mostly taken from Dickson who, not being a doctor, had to take them on trust.

1. Camel's urine: the urine of a she-camel was widely used as a purgative, also as an eye-wash and a wash for wounds. It is still a common hair-wash and they believe it kills head vermin instantly. From our personal observation this cannot be true: perhaps the camels are becoming effete.

2. Kaha or kuhl: an antimony paste which is normally used for blackening and beautifying the eyes, is applied also to the inside of the eyelids as a remedy for conjunctivitis etc. It is supposed to keep off the sun's glare. It continues to be used today. Surprisingly antimony is not used for its emetic and cathartic effects.

3. Girmiz: a pink dye, used by townspeople for sore eyes.

4. Ramram: a wild flowering plant used for sore mouth, tender gums, mouth blisters in children and mouth sores from scurvy, often contracted while pearling. For adults the plant was boiled with water and the resultant thin soup was used as a mouth wash when cool. For children it was crushed into paste with pestle and mortar, mixed with Girmiz and applied to the sore place with a finger. One would like to think that this could have been a source of vitamin C (ascorbic acid) but the heating would probably have destroyed any such value.

5. Al Ja'ada: a wild bush which was boiled in water for use as a purgative. They also used it as a cure for "Basra fever", which presumably, was malaria.

6. Fish soup: was commonly drunk by women living near the sea to increase their flow of milk.

7. Shark's meat: was said to cure impotency in men. What isn't?

8. A donkey's caul was much valued. It was dried and used for the healing of wounds or bringing boils to a head. A piece of the tissue was cut off, soaked in salt water and applied to the part.

9. The common "desert apple" (Citrullus colocynthus) belonging to the colocynth family [3] is still made into an infusion and used as a purge, as its homologue has been in Western medicine until comparatively recently.

b) Branding

The act of branding is still commonly practised today on people of all ages from the new-born to the aged (Fig. 13, 15). A child may arrive at the clinic or hospital in a pitiable condition covered with infected sores: or again it may already be under treatment in hospital but is removed by its parents for branding and then returned for further Western medicine! This is a cruel and horrible practice which has come down in a direct line from the dawn of Arab medicine.

The instruments most commonly used are tent-pins heated to red-heat, but some of the traditional instruments are still in use, the design having remained unchanged for centuries. This barbarous procedure of something that can be seen to be done, is still clung to, perhaps more fiercely as the pattern of life of the Badu changes inexorably from its age-old self-sufficiency to the modern interpendence, and restriction of freedom.

c) The history of Arabian medicine

The centuries before the coming of Western medicine saw little more in the way of "medicine" than what we have just described but it would perhaps be profitable to recall that modern Western medicine owes much to the Arabian tradition, not for its scientific content but because it was such a strong link in the chain of the medical art and observation which must have begun before historical times. It also seems appropriate that the birth of what has come to be known throughout the civilised world for more than ten centuries as Arabian Medicine should have occurred almost within hailing distance of the modern State of Kuwait, which itself has assumed such a leadership among its Arab neighbours in the medicine of today.

But to place our subject in the right perspective we should recall what one of the historians of the Arab people has said [4]: "What we call the 'Arab Civilisation' was Arabian neither in its origins and fundamental structure nor in its principal ethnic aspects. The purely Arabian contribution in it was in the linguistic and to a certain extent in the religious fields".

Like many great conquering nations before and since, the Arabs absorbed and assimilated many features of the cultures of the people they over-ran, and the great civilising benefits which Arab influences exerted to the east and west were extended by people who had indeed the Arab language as their elegant means of communication, but who came of divers nations and races.

d) The Nestorians and Persians

The story begins when we look back over fifteen hundred years, before the rise of Islam, to the founding of a School of Medicine at Edessa in Asia Minor, what is now Urfa in Turkish Kurdistan. The founders were the followers of Nestorius, exiled Patriarch of Constantinople: here was established the earliest centre of Syrian-speaking Christianity. These scholars brought with them priceless Greek texts which were to lay the foundation of Arabian medicine. But continued persecution drove them further eastward and they finally settled at Jundishapur in south-west Persia where they were successful in re-establishing their school. They were helped by the fact that the Sassanian king Chosroes (A. D. 531—581) who ruled from his capital Ctesiphon near modern Baghdad, had refounded the university, which was originally created in A. D. 271 by Shapur I. during which time the teaching of the centuries old Hindu medical theory and techniques had been in Sanskrit. The arch of Ctesiphon remains today to remind us of the greatness of this period of Mesopotamian history, often overlooked. It is of interest that Chosroes's Prime Minister, Perzoes, was also a physician which may explain the support given to the re-establishment of the medical school.

Jundishapur no longer exists but is thought to have been situated between the modern Iranian towns of Shustar and Dizful in Khuzistan. It was here that the Greek texts were translated into Arabic over the next centuries and later into Latin. Thus it was that an heretical Christian sect laid the foundation of Arabian medicine from a synthesis of Hindu and Greek teachings. In the centuries which followed, the term "Arabian physicians" came to be applied to those in many lands who practised the profession and spoke the Arab tongue, though originally Persian, Syrian, African or Spanish. Not all were Moslems, some were Christians, others Jews. In these days of continuing racial and religious tension, it is salutary to recall that one of the greatest 'Arabian' physicians of all, Moses ben Maimon or Maimonides, was not only a Jew, but was persecuted and had to leave Spain because he would not adopt Islam. Notwithstanding he went to Egypt, and later became personal physician to the great Sala'adin in Cairo.

Perhaps the crowning glory of this era was reached with the successive careers of Abu Bakr Muhammad Ibn Zakariyya (A. D. 860—932) commonly known as Rhazes after his birthplace Ray, a village near present day Teheran, Abu Ali al Hussein Ibn Sina (A. D. 960—1037) later known as Avicenna, born near Bokhara in what is now Uzbekistan, a Soviet Central Asian republic, and Maimonides (A. D. 1135—1204). This span of nearly four hundred years, saw the gradual introduction of the Hippocratic tradition into Europe and throughout the vast expanse of the Arab world. Many of the acute observations made by these three physicians and others besides would be acceptable today although their sense of public hygiene might be considered bizarre. It is said of Rhazes that, when he moved to Baghdad and was choosing the site for his hospital, he hung pieces of meat

at various points in the city and selected the place where the putrefaction was longest delayed. Whilst his "Treatment for Snake-bite" and his "Medical Hints for Travellers" have a familiar ring about them, his "Advice on Slave-buying" would no longer be required reading, although one should not dismiss it too quickly for slave-trading has been an unconscionable time a-dying in South-West Asia. Avicenna on his part was one of the world's busiest men, being both a statesman and a physician: perhaps this is why he was also an alcoholic! His "Canon of Medicine" lives after him containing many valuable descriptions of disease and advice on environmental health, which he unfortunately failed to take himself.

In contrast to Arabian medicine which, if it did little good in the curative sense, probably did little harm, Arab surgery was brutal and crude though probably more effective. It is unfair to criticise from a distance of nine hundred years and respect must be paid to that great Spaniard, Abu al Qasin al Zahwari (A. D. 936 to 1018) otherwise known as Albucasis of Cordoba, for he was a man in advance of his time and did much to raise the status of the craft of surgery. Like his medical contemporaries, with the Latin translation of his book "Al-Tasrif Liman 'Ajiza an al-Ta'lif", he brought Arabian surgery to the notice of Europe and with it, high prestige. But the fourteenth century saw the sack of Baghdad by the Mongols and the beginning of the decline of Arabian leadership in the sciences, which was eventually to plunge the areas of Arab influence into a dark age comparable to that of Europe several centuries earlier.

It was not until six hundred years later that the benefits of the new Western medicine founded upon the Arabian tradition, began to flow back into the Arab countries through the media of missionaries, new schools of medicine and large commercial and industrial enterprises instanced in the first place by such early efforts as the East India Company and the Quarantine Service in the Persian Gulf, and latterly by the great oil companies.

Concerning this period of early development, later decline and then resurgence which is now taking place, the historian [4] reminds us: "The phenomenal and almost unparalleled efflorescence of early Islam was due in so small measure to the latent powers of the Bedouins, who, in the words of the Caliph Umar, furnished Islam with its raw material'". The Bedouin (or Badu) live on to furnish the modern Arab world and Kuwait in particular with the essential raw material for industrial progress, man-power.

2. Modern Treatment Services

a) European influence in the Gulf

Arnold Wilson [1], in his book "The Persian Gulf", remarked that there was a very noticeable lack of published information on medical subjects relating to this region and forty years later we cannot disagree with this opinion. During the century following the establishment of Portuguese suzerainty by Albuquerque in 1515 many protective forts were built to guard the new trading posts which had been set up. Two of these were on present day Kuwait's territory, one on the island of Quain in Kuwait Bay opposite Shuwaik, the other on Failaka island. However, no records exist of any humane dealings by the Portuguese with the people of the coastal areas and islands. This is not perhaps surprising because life was a process of the barest survival from day to day and it is doubtful if their ships carried such luxuries as naval surgeons, who might have assisted the needy dwellers in the sparse settlements as was later done by the East India Company and the Royal Navy for over two hundred and fifty years. When first the Dutch and shortly afterwards the English assumed maritime control of the Persian Gulf on receipt of the Charter from James 1st of England in 1616, a thriving trade in the Gulf was already existing between India and Persia. The young East India Company, within a very few years, realised the necessity for establishing some form of disease control in view of the increasing numbers of European merchants trading in the area [2]. In 1620, George Strachan was appointed the doctor at Bandar Abbas in Persia, at the time known as Gombroon. Unfortunately the records of his "irregular behaviour" for which he was dismissed, no longer survive; they would have made interesting reading. He was followed by Thomas Quince who, no doubt like his better known namesake, had to be something of a carpenter too. At that time cholera and plague were the main scourges and continued in that role for the next three centuries, many a promising young factor in the East India Company service being cut down early in his career. The perils and fight for survival in those days must have been appalling yet so little of this comes through in the prosaic despatches to the Courts of Directors in London. In the eighteenth century the East India Company built a hospital first at Bandar Abbas and then, when this was captured by the French, at Bushire. By this time, Kuwait had begun to achieve a reputation for having a comparatively salubrious climate, and short recuperative stays were made there by the Company's staff.

In addition to the climate and the disease, the foreign merchants, Dutch, French and British, had to contend with repeated provocations whether directed at themselves or as part of the constant wars being waged by the Turks against the Persians or vice-versa; one such report to the Court of Directors of the East India Company was made in 1732 by Martin French of the Basra factory. We read that in 1773 Dr. Michael Reilly distinguished himself at Basra, where the Factory had been founded in 1643, by remaining at his post and surviving although some two million of the population were decimated by the plague: on that occasion, Kuwait suffered, the infection being carried there by those fleeing from the Basra pestilence.

It was Kuwait's reputation for health-giving properties together with its natural harbour and gateway to the caravan routes which made the East India Company move the Basra factory there as a result of the factor, Mr. Manesty, having a dispute with the Mutasallum of Basra and the Pasha of Baghdad in 1792. The factory remained in Kuwait from the 30th of April 1793 to the 27th of August 1795, being guarded by sepoys on shore and a British cruiser in the bay; these were necessary because of repeated attacks by the Wahabi sect of fanatic religious zealots.

These dates mark the first known establishment of western medical practice in Kuwait because in 1793 Dr. James Small was appointed factory physician and continued there until the factory's return to Basra two years later. Two peripatetic physicians who visited Kuwait briefly at that time were Dr. Edward Ives [3] and Dr. Seetzen [4], a German physician sent by the Duke of Saxe-Gotha to Levant to collect manuscripts and curiosities. Yet neither felt there was cause either to linger or make any comment, although the former has left an interesting record of the diseases prevalent in the British Naval squadron in the Gulf and the "disorders incidental to Europeans at Gombroon in the Gulf of Persia".

b) The Political Agency and the American Mission

The reputation of British physicians had apparently survived the ill-starred first attempts of George Strachan in 1620 for by the end of the eighteenth century they were in increasing demand both for their professional services and to represent the interests of the East India Company. Kelly [5] referring to the Sultan of Muscat in 1798, tells us that ... "he had lately been reported as desiring to have an English physician by him. Sultan now accepted Bogle (Archibald Bogle of the Bombay establishment) without demur both as private physician and as political agent". It is sad to relate that so many of these servants of "John Company" had too short a life, usually dying by their fortieth year. Bogle survived only two years at Muscat, dying in September 1800, but not before he had carved a small niche in history for himself. It was Bogle who first reported the occupation of the Buraimi oasis in Oman by the Wahabis carried out with the tacit agreement of the ruler of Muscat and Oman. For the next century and a half this fertile oasis was a bone of contention between the neighbouring states assuming even greater importance since the coming of oil because of the paramount need to define boundaries.

The Quarantine Service originally established throughout the Gulf by the East India Company continued to be operated from Bushire until it was transferred to Bahrein in 1928. Although by this time no longer a private organisation, it had had a remarkable record of continuous service and was instrumental in containing many of the frequent epidemics which were brought from India and Africa, in particular small-pox and cholera.

Following the departure from Kuwait in 1795 of Dr. Small, no doctor practised there until the appointment of an assistant surgeon to the newly established British Political Agency in 1904 created by the government of Lord Curzon in India. But, like Bogle in Muscat a century earlier, his duties were not to be solely medical and he was also to have charge of the Post Office; it is of some interest that the first postal mark of Kuwait was franked from the Assistant Surgeon's Office [6].

But the first regular curative medicine ever offered to the people of Kuwait came from a very different direction and was inspired by the clinical success achieved by Dr. Arthur Bennett of the American Arabian Mission of the Dutch Reformed Church in Basra after treating a friend of Sheikh Khazal at Khorram Shahr on the Persian side of the Shatt-el-Arab. Mubarak al Sabah, the Sheikh of Kuwait, hearing of this, invited Bennett to come to Kuwait and so in 1911 Drs. Paul Wilberforce, Paul Harrison, Arthur Bennett and Stanley Mylrea took it in turns from Bahrein to join their lay colleagues of the Mission in Kuwait, founded two years earlier. In the following years Drs. Harrison and Eleanor Calverley were appointed to Kuwait, but shortly afterwards Dr. Harrison was succeeded by one who, for the next forty years, was to claim the hearts of all those in Kuwait who knew him, Dr. Charles Stanley Mylrea.

c) The Government and oil companies

While the work of the "Agency Doctor" as the assistant surgeon at the Political Agency came to be called, remained confined to official clinical and administrative duties carried out in a small office in the Agency building adjacent to the post-office, the doctors of the American Mission were free to grasp the heavy responsibility of bringing modern medicine to as many in Kuwait who wished to seek their services, the success of which was indicated by the fact that at the end of one year they were able to make a start on a hospital building. This was pitiable by the standards in Kuwait today yet was constructed so well with reinforced concrete and steel that it resisted the merciless searing climate for over forty years. Early in the hospital's career its facilities were severely tested when casualties from the historic battle of Jahra in October 1921 were brought to it. Many lives were saved, which earned the lasting gratitude of the Sheikh and people of Kuwait. Hitherto in the history of Kuwait, when battles had been fought men died without recourse to any medical help. By the time the American Mission Hospital closed its doors in May 1967 after a period of fifty-six years, many thousands of patients had been treated and come to depend upon it, so much so that when the decision was reluctantly taken by the Board of the American Reformed Church, there was very strong pressure by the Kuwaitis themselves to retain its services despite what can be described as outstanding government medical facilities available to all in the State of Kuwait. It is fitting that the last Medical Director of the hospital should have been a member of the great medical missionary family of the Scudders who have given so much to medical missionary and educational work in India and the Persian Gulf for a hundred and fifty years.

In the late 1930s the Government of the Emir of Kuwait first considered the need for its own medical service; in 1940 began construction of the Amiri hospital, but it was delayed by war-time problems and not completed until 1949. Meanwhile not more than one government doctor had been available at any one time. In the interim the Kuwait Oil Company, which had found oil in Kuwait in 1936, had begun production in 1946 and in 1947 it erected a tented hospital in Kuwait Town. British, American and Indian medical staff were recruited to run it and it rapidly expanded to become a 106-bedded Nissen-hutted building at Magwa, halfway between Kuwait Town and the oil-field at Burgan forty-five miles away. The Company hospital began to treat the local inhabitants in addition to its own employees and by this time also the State had converted two houses in Kuwait Town; one of these was used as a clinic and the other, in which there was an operating theatre suitable for minor procedures, had a few beds. In 1949 the 80-bedded Amiri hospital was completed and in August of that year Colonel Eric Parry F. R. C. S. formerly of the Indian Medical service was appointed Chief Medical Officer of the State Medical Service. In 1947 Dr. John Guthrie had become Chief Medical Officer of the Kuwait Oil Company and for the next twelve years these two

worked together to improve the medical services of the rapidly growing state. In spite of innumerable difficulties, transport, communications, power supply, inadequate housing, difficulties in obtaining staff for such an apparently unattractive place, the necessity of having to import everything, including drinking water, required for a comprehensive medical service at a time when there was a world shortage, a wide measure of medical aid was available to all by 1955. It had become the foremost medical care offered by any government in the Arab world east of Suez. This consisted of quarantine, school, dental, preventive, veterinary, police and military medical services as well as the general hospital and clinics. Before 1952 there had been no attempt to obtain statistics but then notification of births, deaths and infectious diseases were made compulsory, and in 1959 a Statistical Department was set up.

By 1955 The Kuwait Oil Company had decided that a permanent modern hospital was required for its employees and dependents then numbering about twenty thousand overall, and in 1960 the Southwell Hospital, named after the Company's first Managing Director, was ready for occupation. This hospital, of two hundred beds, has been the envy of all who visit it and it is to the credit of its British architects and Middle East contractors that after ten years it still has the same appearance and functional value for which it was designed. It embodies many of the most modern communication, ventilation and engineering features which have stood the test of time and use. In 1961 the American Independent Oil Company provided a small sixteen-bedded hospital for its staff and late in 1963 the State opened the fine, modern six hundred-bedded Al Sabah hospital to achieve a first class state-wide coverage.

There had also been developed special facilities for tuberculosis, mental and infectious diseases. During these years the population had risen from 120,000 in 1949 to 300,000 by 1961 and has now reached more than half a million at the time of writing. Teaching units for nurses, midwives and health inspectors have been established and Kuwait has been admitted a member of the World Health Organisation which held its regional meeting there in 1964. There are now over three thousand hospital beds (see Text Fig. 23 on the back of Map 3), nearly a thousand qualified nurses and seven hundred doctors and dental surgeons, beside many auxiliaries. However, the service, which is free to all Kuwaitis and makes nominal charges for certain items to non-Kuwaitis, is overweighted on the doctoring side and it is still not possible to provide facilities in the nursing and laboratory fields of the desired standard: not unusual today in other sophisticated countries.

The Kuwait government gives financial and professional assistance to less well-advanced sister states in the Persian Gulf and maintains a permanent medical service for the Haj pilgrimage to Mecca.

d) Professional medical associations

In 1949 the first professional medical group was formed but languished for many years in the face of wider associations of doctors in the Persian Gulf, principally the Persian Gulf Medical Society and the British Medical Association, Middle East branch. However, in 1960 the Kuwait Medical Association was refounded with the backing of its multi-national Arab and non-

Arab members and today maintains a flourishing post-graduate, cultural and social existence and publishes its own journal.

e) Medical facilities and services

Today the widespread nature of the medical facilities offered, from a genetic karyotyping laboratory to comprehensive industrial medical cover, allows Kuwait some justification in its claim that it is among the leaders in the field of comprehensive health care.

From a time when the climate and terrain of this part of the ancient world was a positive hazard for its own inhabitants, marauding armies, innocent travellers and avaricious traders, it has become one of the healthiest parts of the world. On both sides of the mouth of the Shatt-el-Arab rapid development is taking place and when the Kuwait Medical School is founded in the near future, this will complete a geographically related trio of medical schools, for Basrah has been established now for some years, and the medical school of Ahwaz in South Western Iran has been refounded almost on the historical site of the ancient school of Jundeshapur. But until now all medical staff in Kuwait has been trained elsewhere and virtually all the nursing staff, except for the past four years when the class of 1964—unfortunately only a handful—graduated in 1967 at the Al-Sabah. Elsewhere Nursing Assistant training is given (the Southwell Hospital) and gradually paramedical training is being introduced, although for some years this has been reasonably advanced in the field of public health inspectors.

The structure of the treatment, as distinct from preventive services is identified from the accompanying "family tree" (see Table 32 on p. 58). Much care and planning has been given to the regional siting of the diagnostic and treatment clinics together with the establishment of a medical record system and the identification of the urban and rural groups of people with their nearest clinics. Whereas these latter are open during working days, central clinics are available on a 24-hour schedule and serve a wider area. While the intention is sound nevertheless the rules are not adhered to by a population as emotionally erratic as are found today in Kuwait, and, as in many other countries, the example is not set by those who should know better, for many well-to-do-people attend specialist clinics in the government hospitals without being referred by a general practitioner. This has interfered greatly with the development of good after-care services and the communications between the several doctors who may have been consulted by the patient; in the end it is the patient who fails to reap the full benefit. But while the overall operating efficiency of the clinics is prevented from being fully developed due to improper use, nevertheless, the medical coverage given to anyone is rapid and reasonably comprehensive for those who do use the government service today in Kuwait. One feels that this may not have been so, in the recent past for those people whose diagnosis and treatment lay in the hands of private practitioners, not all of whom have practised either effectively or ethically. This has led the government to impose restrictions and control on those entering private practice and in particular when seeking to open an in-patient clinic. The results of this government policy have been very satisfactory in achieving its aim and the overall quality of private practice

today compares well with that in other sophisticated cities of the world.

The staffing of the government clinics has presented a problem because sufficiently well trained auxiliary staff has been hard to come by and the great weight of patient attendances serves to deflate the enthusiasm of most doctors after only a short exposure. The policy of cycling the medical staff through clinics and hospitals alternately is in theory the one practised, but in fact, there is considerable reluctance of many junior staff to return and of their seniors to let them return to clinic practice. This has had the result that over the years the clinics have become the entrepot for the patients, which is as it should be, and the final repository for many an ageing and disillusioned physician. That the authorities are aware of this is shown by the considerable efforts made to include peripheral clinic doctors within a regular framework of post-graduate education. This itself is uphill work and the response has not been rewarding; yet it is not a phenomenon unique to Kuwait as secretaries to local medical societies the world over know to their chagrin.

With the high ratio of cars to households and a satisfactory public transport system, no one in Kuwait is without medical aid except for those deep in the desert —even here the transport camel is fast giving way to the motor truck which brings the badu to within a few hours at most of assistance.

Everyone is entitled to see a doctor at the clinics on request; very often this is abused by a demand for the whole family to be examined. The concept of paramedical consultations is not accepted in Kuwait and indeed it could be argued as unnecessary in view of the high doctor-population ratio, one doctor to every 750 people. Nevertheless the doctor is often burdened with trivialities, and this together with the very high attendance rate leads to enormous clinics, allowing time only for the minimum of counselling and the briefest of examinations. The radio and television health education programmes are increasingly listened to with the result that clinic attendances are further encouraged and the demands of the patients can only be met by further hospital consultation. It is very probable that as the general education improves there will be a change in this attitude, but there is no doubt that all the present services have in fact created a demand with which they are struggling to cope. Added to this is the fact that all clinic services and hospital outpatient services are free, thus affording no brake. One of the many problems is that a patient may be treated for the same illness with a variety of drugs prescribed by different doctors and without each other's knowledge. The patient may benefit, but often the illness is complicated by overtreatment and interaction of drugs.

Undoubtedly the value of the clinics is greatly enhanced by an efficient transportation system: an ambulance is available at all times by radio contact to transport sick to hospital and this service has contributed greatly to the lowering of mortality figures.

The general hospital system has steadily developed since the first government hospital, the 80-bedded Amiri, was opened over twenty years ago. Relative to the steady expansion of the population, Kuwait today must have one of the most effectively organised hospital systems in the world; this is shown in the Text Fig. 23 back of Map 3. It should be pointed out that since 1964 scales of payment for hospital care have been applied to non-Kuwaitis.

Perhaps it is justifiable in retrospect to level criticism at the original planning by pointing out the direction in which the available money has been misapplied, if only to assist planners elsewhere. This is not to impute either negligence or even ignorance of modern requirements but rather to show that demands upon medical man and woman-power the world over have exceeded the available supply on an almost logarithmic scale, but realisation of this has been slow. Over the past twenty years it has been possible to staff the Kuwait hospitals with nurses, doctors and technical staff, but only at the expense of other countries in the Middle East, India and Pakistan. The provision of sophisticated medical services on the scale of Kuwait has required professional and technical assistance, supply of which continues to be a far more difficult task than that of buildings, engineering services and equipment, difficult as some of those have been. They are illustrated by the problem encountered with the high water-table when building the Al-Sabah and new maternity hospitals.

Recently however these manning and physical problems have been approached together and considerable care has been taken to tackle them in parallel. Dr. Birt Lindgren, the Swedish hospital architect has prepared extensive plans both for the further development of the Al-Sabah hospital and also for the as yet unbuilt 1200 bed Mubarak hospital. The nursing training is being expanded, several hundred medical and medical-technical students are training outside Kuwait in the Middle East, Britain, Europe and North America. As they return they should find facilities second to none in the Middle East. But still the greatest obligation will be to persuade Kuwaiti girls to enter nursing, and for some time to come reliance will have to be placed upon neighbouring Arab countries. In the early years of the Kuwait medical service assistance was obtained from nurses from India but for a variety of reasons, not least competition from elsewhere and restriction within India itself, this source is now limited. In the future more emphasis will have to be given to human and mechanical nursing aids and the concentration of trained staff to care for those sick people whose illness justifies highly qualified nursing: Kuwait shares this problem with other countries where the development of techniques is having to adjust to the competition for young man and woman-power to enter fields other than the nursing and medical-technical professions.

The government has long taken the attitude that both general and special facilities should be developed concurrently and this has been followed by beneficial results, particularly in the fields of tuberculosis, psychiatry and orthopaedics. The relative priorities have varied with the years but with a population which is now living longer, and thus developing malignant disease, expansion of the limited radiotherapy services has been ear-marked. During the past fifteen years many people from Kuwait, both Kuwaitis and others, have been sent abroad at government expense for such treatment, mostly to the United Kingdom. This is also the case with neurosurgery and other ultra-specialist needs, but with the provision of facilities in Kuwait the need for this should no longer exist: no doubt pressure will still be brought to bear on the government by influential people for these privileges to continue. The wisdom of a small country arming itself with these expensive paraphernalia of modern medicine is beyond this present discussion but should perhaps be viewed on a regional rather than a national basis.

The quality of buildings, fittings and technical equipment is of the highest and again reflects the desire of the Kuwait government to provide only the best available: nevertheless this policy has created great demands upon the maintenance staff and the various servicing organisations set up in Kuwait. It has been possible to estimate the relative national marketing endeavours of the great exporting countries by the ability of the competing firms, American, European and Japanese to provide an effective and prompt service. Deterioration of electronic and other delicate equipment is rapid in such a country as Kuwait. It is therefore imperative that every encouragement should be given to firms supplying equipment to establish efficient servicing. It is noteworthy that some countries are considerably more ready to do this than others, taking the view that while the expense may eat into short-term profits, the benefit reaped over the long-term will amply repay the outlay.

f) Laboratory services

Two auxiliary services are of particular importance in supporting a modern hospital organisation, one being the over-all laboratory facilities, the other an efficient blood bank. The former has been bedevilled by the difficulty in finding qualified staff, from pathologists to junior technicians, and this probably follows the pattern of the whole of the Middle East where, but for a few exceptions, laboratory services are either deficient or unreliable. This itself generates trouble far beyond the immediate problem because generations of doctors are trained who come to practice their profession without the help of these important tools of medicine, are not encouraged to enter the speciality and find too often that clinical satisfaction can apparently come their way without the need to submit specimens for testing.

The above strictures however cannot be levied at the medical services of the Kuwait Oil Company where, for over twenty years, first-class laboratory in microbiology (without virology), biochemistry and haematology have been carried out until recently with the assistance of British and Indian staff. Histology has followed a varied career however and for the past ten years it has been found expedient to send the prepared slides to a consulting pathologist in the United Kingdom. Urgent reports have been relayed to Kuwait within 48 hours and the service has proven very satisfactory. The government has also followed this practice intermittently depending upon the presence on their staff of a morbid histologist. Kuwait however has failed to offer to any morbid anatomist that feature of his work which is so essential for professional expertise and satisfaction, autopsy material. Autopsies have been proscribed by government policy and Moslem religious restrictions, with the exception of those few non-Moslem patients whose relatives have permitted examination. Within the past few years the government has modified its stand and has made it mandatory that autopsy should be carried out on all accident victims or those dead of suspected foul play, but only the government "forensic expert" conducts these examinations and they do not have the more general educational value of routine autopsy material.

g) Tuberculosis diagnostic laboratory

In 1961 the W.H.O. tuberculosis expert Dr. Jensen set up a tuberculosis diagnostic laboratory for the government, the only culture and sensitivity testing up till that time being done in the Kuwait Oil Company hospital. Following the establishment of the laboratory and the training of staff, by 1962 it was possible to concentrate all tuberculosis culture and sensitivity tests in the one laboratory. The example and inspiration of this effort encouraged the development of more wide-spread bacteriology investigation until at this time it is on a solid foundation, although staff to run the service remains a continuing problem.

For many years the Kuwait Oil Company hospital laboratory assisted the government service and both continue to make use of the special liaison with the United Kingdom Public Health Laboratory Service at Colindale—particularly with regard to the typing of organisms and identification of virus antibodies and culture. On occasion material has also been examined by the Virus Institute of Moscow, U.S.S.R.

Within the past three years however an effort has been made to repair the past Laboratory staff deficiencies with the help of overseas training programmes for young Kuwaitis, experts from outside Kuwait and the provision of good facilities and equipment. These have indeed been demanded by the well-qualified doctors from many countries undertaking investigation and research. Results are now seen by the steady flow of useful clinical, pathological and epidemiological information.

One of the most difficult problems in setting up services requiring technicians is the status of such people: Kuwait is not alone in this for in Britain also the question of the non-medical scientist is being actively discussed. A compromise has in fact been reached already in Kuwait in certain fields and the pharmacists now enjoy equal professional status.

h) Blood transfusion services

Blood transfusion has been practised in Kuwait for fifty years without any religious, social or other restriction but until 1965 only the Kuwait Oil Company provided its own emergency blood bank. All donor blood prior to this date was either collected immediately before use or obtained from overseas, Cairo, Lebanon, the U.S.A. and elsewhere. The U.S.A. has provided a reliable if expensive supply, which was effectively developed by the Kuwait Oil Company initially, but it requires very careful maintenance of blood donor pack temperatures en route and immediately after arrival in Kuwait by a foolproof systematic scheme. To avoid reliance entirely on these remote sources the Government of Kuwait established a Blood Bank in May 1965 [7], gradually changing from the supply of imported blood to almost complete reliance upon local donors by 1969. In 1965, 2480 donors were bled, rising to 6983 in 1968; this has undoubtedly been assisted by the fee of K.D.10 paid for each donation to residents of Kuwait, although only K.D.5 is offered to visiting seamen at the ports. The main blood collecting centre is in the central laboratory of the Amiri hospital and there are two branches in the Ahmadi area. The collected blood is issued to all hospitals except the Southwell hospital, which maintains its own local and imported supplies.

Thus a very fair balance is being struck. A great strain was put upon the service in June 1967 at the start of the Arab-Israeli war when thousands of donations of blood were taken and shipped to the Suez Canal areas for use by the Arab army hospitals. The blood bank has endorsed a regular programme in blood bank technology for all technicians in the Central Laboratories, to obtain their help in case of a national emergency. A qualified technician is now on a training mission in the United Kingdom for plasma lyophilisation and extraction of blood coagulation factors.

The establishment of the blood bank provided the opportunity to examine the prevalence of the major (A B 0) blood groups among 10,000 donors comprising the different ethnic divisions of the Arab population in Kuwait [8]. The gene frequencies identified in Kuwaitis and among the tribes of Arabia (Table 40) support what is known of the tribal migrations to Kuwait, being closest to the frequencies reported from the Badu of the Nejd and Southern Iraq (Text Fig. 24).

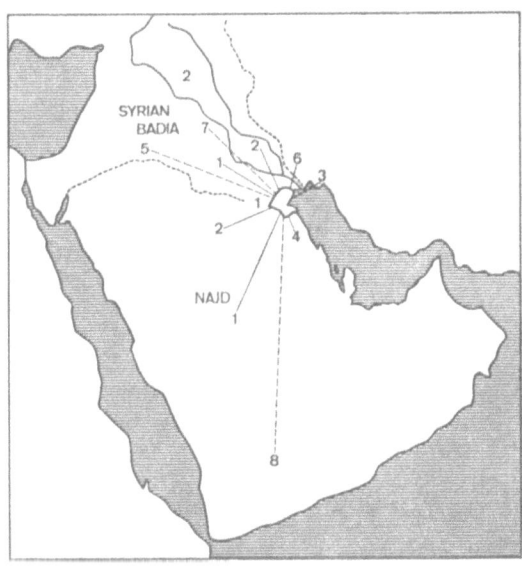

Text Fig. 24. Sites of tribes. The sites where the main tribes, that had contributed to the Kuwait population, live.

1 Anaiza tribe	4 Awazem	7 Akeydat
2 Shammar	5 Rowala (Anaiza)	8 Southwest Arabia
3 Bani Ka'b	6 Maualy	

Solid lines: ancestral relationship verified historically
Interrupted lines: possible relationship suggested in this study

The integration of the hospital services within the general framework of preventive health services and pathology institutes is an essential prerequisite for the knowledge and effective control of disease within a country's borders. A case can also be made for close co-operation with similar services in adjacent political units because, with their common geographical factors yet differing economic and social patterns this could emphasise the differences in disease characters and give a better appreciation of their causes, and encourage concerted action for their disappearance or suppression. Until now this has had to rely upon loose associations under the auspices of the World Health Organization and associated organizations, rather than close co-operation between neighbouring governments.

Table 40. *The A B 0 gene frequencies of Kuwaitis and those Arab tribes with high 0 frequencies*

| | Group percentages | | | | Gene frequency | | |
	A	B	AB	0	A	B	0
Kuwaitis	24.24	24.13	4.37	47.26	15.52	15.42	69.06
Nejd Arab	24.08	29.23	3.08	44.61	14.20	17.82	67.98
Badu near Baghdad	26.63	25.74	6.80	40.83	18.43	17.89	63.68
Rowala tribe	11.25	5.62	0.94	82.19	6.3	3.3	90.4
Manaly tribe	7.51	3.29	0.00	89.20	3.83	1.66	94.51
Akeydat tribe	16.98	5.66	2.52	74.84	10.23	4.16	85.61

i) Relative paucity of published clinical material until 1968

While the relative paucity of clinical material reported from the Kuwait hospitals may be justified by the absence of statistics and laboratory support, the material available from the preventive field units contrasts significantly. Evidence is available to show that this trend is now being reversed but this will be slow and the effects uneven unless certain needs are provided:

1. Moral, intellectual, practical and financial encouragement from all levels of the population. Here the question of autopsy material is paramount.

2. Leadership in research from Ministry level to the smallest unit.

3. Library services of a high standard.

4. Continuing post-graduate education both within and beyond the borders of Kuwait.

5. Provision of supporting paramedical workers, —including nurses—on an increasing scale and the recognition of their contribution by an accompanying improvement of status.

6. Increasing co-ordination of health education with the available preventive and treatment services in Kuwait.

X. Trauma, Temperance, Tuberculosis and Toxoplasmosis

1. Trauma

Traumatic death and injury are nothing new to the Arab of Arabia, despite the alarming figures of automobile accidents which are increasing yearly.

a) War and wounds

When *the last great battle* between the Kuwaitis and the fanatical Wahabi Ikhwan tribesmen of the army of King Ibn Saud was fought fifty years ago, *in October 1920*, for the first time the wounded received modern medical treatment at the hands of the American Mission Hospital in Kuwait. Dr. Stanley Mylrea, who was in charge of the hospital, has remarked that "There can have been few more bloody fights in the history of Arabia. The Ikhwan, who went into action with some 3,500 men, lost about 800 killed and as many more wounded, nearly half their total strength. Their wounded, with no skilled medical attention, died like flies ... in numbers the two armies were fairly evenly matched, but the Kuwait men were careful to take cover while the Ikhwan scorned to do so ... of Kuwait forces only 63 were killed in action, four more died of their wounds in the American Mission Hospital. Kuwait had about 120 wounded. Of the four Kuwait wounded who died, one was a case of gas gangrene, the only case of its kind I have seen in Arabia. The battle of Jahrah was fought in heavily manured date gardens and grain fields. There followed a high incidence of infection ... The injuries were of course various, some slight and some grave. There were a few sword and dagger wounds but nearly all were gunshot wounds." [1]

Fifty years on and the Kuwait Arabs have quickly developed a daily toll of death on their modern roads which makes their earlier losses from tribal and religious wars pale into comparative insignificance.

b) Deaths and accidents on the roads

It has been estimated that there is approximately one private car for every four persons in Kuwait, but this figure must be treated with reserve in view of the other types of vehicles used by families and the rapid resale of cars. With only 1,200 kilometres of roads to accommodate over 150,000 vehicles of all types this gives a figure of 125 to every kilometre or each vehicle occupying 8 metres of road space if every vehicle was on the road simultaneously—at times this seems to be the case. In relation to road accident trauma the accompanying figures illustrate a dangerous trend in motorcycles but one mirrored in other countries with a teenage bulge (Table 41). With increasing numbers of cars registered and the sharp rise in motor-cycles of all kinds a concomitant very sharp rise in injuries and deaths has occurred (Table 42). This in itself should be a lesson but as yet no legislation to limit motor-cycles has been enacted and for a number of reasons is not likely to be; only education in the use of vehicles and highway code is going to reduce these serious figures.

c) Statistics

The rise in traffic accidents has unfortunately been paralleled by a similar rate increase in *industrial and house accidents;* great efforts have been made by govern-

Table 41. *Road accidents resulting injuries and deaths.*
1958—1967

Period	No. deaths	No. injuries	No. accidents
1958	81	1159	3564
1959	91	1535	4502
1960	103	1911	5406
1961	98	1962	4172
1962	95	1442	5185
1963	115	1758	6739
1964	117	1686	6036
1965	123	1622	6555
1966	198	1653	8588
1967	213	2231	10203

Table 42. *Rise in road accidents*

	1960	1961	1962	1963	1964	1965	1966	1967
Motor cycles	180	318	222	3988	4517	4791	4076	1073
Accidents	5406	4172	5185	6739	6036	6555	8588	10203
Injuries	1911	1962	1442	1758	1686	1622	1653	2231
Deaths	103	98	95	115	117	123	198	213

ment and other organisations to educate the public in preventive measures, but so long as overcrowding exists so will domestic hazards. One of the most serious, and recognised universally wherever it is used for domestic purpose, is the accidental swallowing of kerosene by very young children, who go to a tap, unfortunately too often the tap of the kerosene container, to obtain a drink.

No specific accident data have been collected over the country as a whole, but Thom [2] carried out a survey among the 15,000 family members (not including the father) in the care of the Kuwait Oil Company in 1965 and this gave an overall rate of just over 14 per cent (see Table 43). The rate of accidents dropped significantly during the cool weather (November, December, January, February), except for the "paint effects" in the early part of the year, explained by the introduction of

an inside wall paint for houses which contained a fungicide which vapourised with the heating and closure of windows in the cool weather, but was apparently effectively dealth with by through ventilation, fans and air-conditioning in the summer. However, the largest single factors were traffic accidents and burns; the latter being due to matches, kerosene and scalding by tea, coffee and hot water. The majority of accidents occurred in children and these are worth a closer look (Table 44).

The cause of the high rate in Arab children can only be assumed, probably relating to size of family, social circumstances, cultural habits and location of housing in relation to traffic concentration. The first-aid, ambulance and hospital facilities were equally available to all four groups.

No comparable figures are available from Government sources but there is no reason to suppose that the circumstances differ very greatly except perhaps in the time taken to reach hospital. Of the 1,904 Oil Company childrens' accidents, 76 required admission to hospital and there were two deaths.

A major influence which has unfortunately contributed to an increasing toll of traffic accidents has been the imposition of alcoholic drinks prohibition which came into effect in 1964. Many tragic accidents have occurred on the return journey from Basra, just across the border in Iraq, after an evening or weekend's relief from restrictions. This is doubly distressing for the prohibition law was designed, along with other apparently good reasons, to reduce the increasing incidence of mutilating and fatal automobile accidents.

2. Temperance

Indulgence in fermented liquors runs contrary to the tenets of the Holy Koran, although their actual interpretation has differed among zealots and the more liberal minded since the birth of the Mohammedan religion in 622 A.D. Throughout Moslem countries *religious law* and its common observance may differ in varying manner; in several countries wine is universally produced

Table 43. *Accidents in family members*

	Jan.	Feb.	Mar.	Apr.	May	June	July	Aug.	Sept.	Oct.	Nov.	Dec.	Total
Burns	10	14	20	9	9	20	14	16	18	23	11	20	184
Fractures	6	3	14	9	3	4	15	7	13	13	4	5	96
Foreign bodies	2	2	3	10	9	16	10	9	7	6	2	10	86
Dog-bite	1	—	—	—	1	—	—	—	—	—	—	—	2
Paint effects	—	10	18	18	—	3	4	1	10	9	4	3	86
Kerosene poisoning	4	1	2	3	1	1	1	3	—	—	—	—	16
Other and minor	108	81	135	153	124	146	185	134	178	191	116	98	1649
Total	137	111	192	202	146	190	229	171	226	242	139	138	2119

Table 44. *Accidents in children*

Nationality	British			Kuwaitis			Other Arabs			Indians/Pakistanis		
Age group	0—2	3—5	6—12	0—2	3—5	6—12	0—2	3—5	6—12	0—2	3—5	6—12
No. at risk	100	200	600	650	740	1817	850	1100	750	600	1050	750
Accidents	13	27	53	206	314	280	226	329	165	72	104	115
Total		93			800			720			291	
Rate/100		10.3			31			37.5			41	

and drunk, such as Algeria, Egypt, Turkey, Lebanon, Iran, while beer is brewed in Iraq. Spirits are also distilled and usually marketed in the form of Arak, the very stimulating aniseed flavoured drink. Until 1964 all alcoholic drinks were imported into Kuwait and the Government received a respectable sum in customs dues, amounting to nearly 1 million K.D. in 1963 (actual value of import 558,828 K.D.). After this date, import ceased except for the diplomatic missions, the value in 1967 being only 36,963 K.D.

It would be unrealistic to say that alcoholic liquor is not made in Kuwait today, for newspapers frequently report the discovery of a still used to produce liquor for resale; very heavy penalties exist for this crime. On the other hand the brewing of beer, the making of wine, and the distilling of spirits for personal consumption is permitted to non-Moslems, who must consume them in private. The opportunity and temptation to outwit the law is considerable for there is a ready sale to people in Kuwait not withstanding the religious objection. Occasionally caravans of liquor are confiscated on the border of Kuwait and Iraq and boatloads from Iran and visiting ships have the same experience. The profits of smuggling branded liquors are high and events elsewhere should have convinced the Government that *prohibition* only drives people to clandestine production and importation. The enactment of prohibition by the Majlis (House of Assembly) was influenced by a number of factors in addition to the strong lobbying for the acceptance of Koranic ruling by the more orthodox representatives. Perhaps the most controversial was the fact that all importation was in the hands of one expatriate firm who had received the franchise many years before and it was felt that this was unfair once profits began to reach such high levels. Another reason was the growing awareness that increasing numbers of young Kuwaitis were becoming addicted to alcohol. Unfortunately prohibition is no answer to this problem: Many of these young people continue to obtain supplies and the remarks earlier about the traffic accidents on the road from Basra apply particularly to them. That the Moslem Arab and expatriates do indulge is confirmed by Salem's figures [3], obtained during the first fourteen months after prohibition. Over that period 168 people were admitted to Kuwait Hospital with acute alcohol intoxication, 114 from ethyl alcohol (C_2H_5OH) and 54 from methyl alcohol (CH_3OH). There were 37 deaths, mostly due to methyl alcohol and seven patients had permanent blindness.

Current figures are not available but the frequency is now thought to be diminishing: This could be explained in two ways—either consumption of both alcohols was decreasing or, and more likely, supplies of better quality ethyl alcohol were reaching the consumer and reducing the resort to methyl alcohol. It is interesting to note that the import "manufactured perfumes, alcohol" rose to a high figure of 194,305 K.D. in 1966 and has since declined; it is in the form of alcoholic toilet preparations that many people have satisfied their needs in the first years of prohibition. Further evidence of the continued use of alcohol is that there has been no rise in other forms of intoxication and addiction such as hashish, kat and modern drugs following prohibition, which might have been expected.

Conflicting impressions are held by those within and without Kuwait concerning the granting of liquor licences to visitors. Despite suggestions in Government tourist literature that liquor may be officially obtained in Kuwait, this is definitely not the case, but as the old saying goes, "There is no harm in trying"; tea and coffee-pots have seen strange bedfellows.

One major drawback of a prohibition programme is that consumption is driven underground and attempts to identify and treat all forms of alcoholism are frustrated by the fear of condemnation, whether legally, from employers or society.

Finally, to emphasize again the close connection between temperance and trauma, presence of a spirit-still in a house may constitute a great danger; many a home in Kuwait has caught fire since 1964 and many an individual seriously injured. Figures associating these accidents with alcohol distillation however, are lacking for obvious reasons. Not so those of the loss of Kuwait currency to Iraq through weekend consumption of alcohol in Basra, estimated to be in excess of six million K.D. yearly!

3. Tuberculosis

Historically, tuberculosis and small-pox are probably best documented; certainly we know that both have existed with some frequency since the second millenium B.C. and the descendents of the original desert dwellers with their urban counterparts still suffer the same fate given the environmental conditions, and perhaps to a slightly greater extent than might be expected based on figures from western countries, although consistent with experience in south-east Asia and Africa. The traditional friend of man in the desert, the camel, has the same proclivity. Dickson [4], who had no rival in his knowledge of Kuwaiti tribes and their neighbours and who was not a doctor, observed that "... consumption is primarily due, I think, to malnutrition, by sharing the same coffee cup with others and the prevailing habit of covering the head completely with the quilt or lahaf when sleeping at night in winter. The Arab has the further unhealthy habit of sleeping under the same lahaf as his wife, each breathing the other's breath throughout the night ... another possible cause of consumption may be the quantity of dust and sand the average man and woman breathes during the course of a year, especially during the summer months, when by day and night dust lies heavy over the camps and sandstorms are of almost daily occurrence."

It is worth quoting Dickson at length to be able to contrast the enormous improvement in the chances of recovery and ease of treatment which has taken place in Kuwait in less than twenty years. "As soon as a person develops a cough with continuous low fever (it is curious how much more often women and girls seem to get the disease than men), the relatives take the patient and place two straight brands on the left wrist ... if this does not bring about an improvement in fifteen days or so, two more brands, similar to the above, are placed on the right wrist. Lastly, if no improvement is seen the bottom of the tongue is similarly branded. This operation is terribly painful, it causes the mouth to swell up, and for days the patient cannot speak at all. After this no more is done for the affected person. If she or he dies, it is maktub'alai (it is written his time is come)." Today early diagnosis, treatment by specific drugs, good food, a short hospital stay and careful follow-up have contributed to a rapid abatement of the individual's distress and danger to the community.

A feature of the epidemiology which Kuwait appears to share with other developed countries experiencing increasing longevity of their people is the reactivation of quiescent *tuberculosis in the older person,* and subsequent infection of the young. Not confined to Kuwaiti Arabs alone, this phenomenon is a worrying one where antituberculous vaccination of babies with B.C.G. (Bacille-Calmette-Guerin) has not been universal. The cultural habit of the Arab to employ elderly relations as helpers in the house and to care for the many children exposes the latter to grave risk of infection.

The random *sample of population* reported in 1962 [5] during a Government survey gave a prevalence rate of 3.6 per thousand which reached its highest level among Badu; not unexpected. However, the progress in hunting down this ubiquitous disease has been very successful although up to 1966 the incidence rate of inception of tuberculosis still rated highest of the most common serious infections.

In 1960 a ward of 20 beds was opened for tuberculous patients at the Amiri Hospital, but the demands far outstripped anticipated use and rapid *expansion of facilities* had to follow. This pressure was occasioned by the unrestricted admission of all who were diagnosed in Kuwait, many coming to Kuwait for the express purpose of being treated. Expansion continued at a high rate until 1966 when control of visitors was imposed; in fact a large new chest hospital had already been built only to find itself redundant before it was commissioned. It is now the main maternity hospital in Kuwait, for at the same time demands for beds of this type arose; nevertheless, as can be imagined, considerable internal reconstruction was required.

State services for tuberculosis have now become organised under the following scheme and diagnosis reaches far out into the desert through the efforts of the Preventive Unit. The chest diseases department comprises:
1 — Preventive unit
2 — Central chest clinic
3 — Two sanatoria (total capacity = 675 beds)
4 — Central T.B. laboratory
5 — Chest surgical unit
6 — Physiotherapy section
7 — Occupational therapy unit.

With over 70,000 outpatient visits a year and 1,200 new cases in 1966 the problems of control are still not solved. However, many of these patients came from neighbouring countries and were not permanent residents. The bed accommodation has increased from 20 to 675 over a period of 15 years and full surgical facilities are available for pulmonary disease and closed heart surgery not requiring a heart-lung machine.

In 1960—61 a World Health Organisation team visited Kuwait and installed a modern diagnostic and drug sensitivity-testing laboratory. Whether this has influenced the comparatively low rate of drug-resistant infections is not clear; probably the coincidental development of good follow-up services has helped to control the fall-out from treatment, known to be a potent cause of drug-resistance. *B.C.G. vaccination* is carried out principally on students; between 1962 and 1964 66,832 students were tuberculin tested and 18,972 on being found negative, were vaccinated. Ideally, new-born infants should be vaccinated but in an open community there are almost insuperable obstacles to this at the present time. The benefit of those actually performed is dimin-

ished by the low frequency of return for tuberculin testing afterwards, which is such a valuable tool for the epidemiologist. In the Kuwait Oil Company population it has been possible to identify the value of the procedure by post-testing, although initially this was hampered by using only a 1 Tuberculin Unit strength which gave a low percentage of conversions in the very young [2] (Table 45, 46).

Table 45. *Tuberculin test in newborn and children*

Age	Number	Tuberculin test (Mantoux negative)	Conversions per cent
New-born	241	169	30
1	154	90	42
2	141	64	55
3	100	51	49
4	124	59	52
5	90	42	54
Total 0—5 years	850	475	44
6	94	40	57
7	101	32	68
8	58	17	70
9	56	21	61
10	50	22	56
11	12	5	60
Total 6—11 years	371	137	64

Table 46. *Tuberculin test in children*

Age groups	Examined	Negative	Percentage negative	Percentage positive
5—6	82	69	84	16
6—7	85	76	89	11
7—8	44	30	68	32
8—9	35	23	68	32
9—10	46	28	60	40
10—12	55	30	54	46
Total	347	256	74	26

This low *conversion rate* in the very young possibly reflects the diminished capacity of this age group to develop allergy although a good tuberculin response appears immediately a natural infection occurs. In the following year, 1966, using a stronger tuberculin strength, 5 units, some 80% of babies showed a good B.C.G. scar and conversion to a positive tuberculin test [6].

Nevertheless, there is no room for complacency because there is evidence that infection in certain groups of children may be increasing; in 1961 Indian and Pakistani children in the age group 5—9 had 15.5 per cent positive on initial tuberculin testing with 1 Tuberculin Unit. Five years later in 1966 this figure had increased to 19.5 per cent and was accompanied by evidence that two out of every three children presenting themselves at school (the site of the testing) had not had protection by B.C.G.

The experience with this group of children emphasised the hazard to others who may enter the country with their parents without having received B.C.G. or have been born within the country and escaped B.C.G. vaccination. It has been the practice in this Oil Company population to give chemotherapy prophylaxis to all positive tuberculin reactors under the age of five years because of the evidence of their recent infection and to give B.C.G. to all tuberculin negative children and adolescents. It is agreed that this is a policy of idealism which can seldom be practised unless the community is under close supervision: Unfortunately this is not the case elsewhere in Kuwait at present, but may become so when medical records are standardised and available for every individual. Computerisation will assist in reaching this admirable goal, but illegal immigration will have to be controlled first.

Finally a word about *non-pulmonary tuberculosis* which, interestingly, does not occupy quite the same relation to the pulmonary form as has been seen in western countries over the past seventy years. Booz and El-Gayar [7] reviewed 216 patients with non-pulmonary tuberculosis who presented during 1964 with skeletal disease. They confirmed Dickson's original observations of the proneness of the Badu and preponderance in females and reached the same conclusions as to the cause. There was a high incidence of associated lesions, skeletal and otherwise, but tuberculosis of the urinary tract was not seen; it is very rare in Kuwait. Apart from bone disease very little non-pulmonary disease is seen among the indigenous peoples, although this is not true of expatriates, particularly those from the Indian sub-continent.

Thus Kuwait presents a commendable example of effective public health control and organisation for treatment, yet with the shadow of the ever present danger from across its borders of a disease probably as old as man himself.

4. Toxoplasmosis

In biological terms we are justified in regarding Toxoplasma gondii and the human body as having existed in symbiosis over a much greater period of man's existence than has the more familiar tubercle bacillus. To further emphasize the palaeozoic priority of T. gondii, it has the same symbiotic relation to many species of warm-blooded animals, mammals and birds, and indeed even to a few cold-blooded reptiles. Within the last few years there have been suggestions, together with some experimental evidence, that invertebrate hosts may contribute to the transmission cycle of the human and animal disease thus affirming its precedence.

Until recently this protozoan parasite apparently exhibited an epidemiology almost entirely confined to temperate and tropical areas of the world which possessed a significant rainfall. The heaviest prevalence was in low-lying, moist littoral with some unusual peaks in such widely discrepant areas as Easter Island, latitude 109° W. longitude 27° S; New Zealand 170—180° E. 37—50° S; Tristan da Cunha 12° W. 37° S; and the James Bay area of Northern Canada 80° W. 52° N. The fact that in these areas there is extensive association of man with domestic or grazing animals or wild animal trapping suggested that there were factors other than climatic or geographical which might give rise to the pathogenic mutation of the organism. It was felt that perhaps infection by other agents, virus, bacterial, protozoal or metazoal could interfere with existing immunity mechanisms.

By the early years of the last decade an impasse seemed to have been reached, for no concrete evidence

of the transmission cycle existed. It was felt justified therefore to review the possibility of the infection occurring in Kuwait in view of the close association of many of its people with animals, particularly sheep and goats, the former being prey to enzootic infection in many parts of the world.

A study was undertaken [8] to examine the prevalence of human and animal infection, both latent and active. It occupied six years from 1962 to 1968 and comprised two stages, a pilot survey of sheep and goats with testing for Toxoplasma antibody in the blood, and human case-searching.

Significant levels of antibody were found in Kuwaiti sheep comparable to a control group of imported Australian sheep from a known Toxoplasmosis endemic area in Queensland (Fig. 55) and some 60 per cent of goats tested showed antibodies: On the other hand goats which had been isolated since they were kids showed no evidence of infection.

Table 47. *Age distribution and prevalence rates in Toxoplasmosis*

Age group	Population	Proportion	Toxoplasmosis cases	Rate per 100,000
0—10	6000	33%	5	83
10—20	4000	20%	2	50
20—30	2000	9%	5	250
30—40	3500	19%	6	171
+40	3500	19%	3	86

In the human series, 21 patients satisfied the clinical and serological criteria adhered to in a previous study elsewhere [9] (Table 47). The age range reflected the general pattern of the population, 2 to 41, but there was an unexpected preponderance in the 20 to 40 group. The predominant picture was a generalised enlargement of the lymph nodes; two of the patients had concomitant tuberculosis and were Omanis; three patients had enlarged spleens and one of these had a rise and fall of Brucella antibodies without specific treatment. The blood of three patients, all of them with lymphadenopathy, showed a fall of the granulated white cells, Eosinophilia was occasionally seen; its significance was related to accompanying worm infestation, whose presence could possibly be contributory to the toxoplasma infection. The interesting observation was made that, whereas in the West African series [9] five of the 63 patients or 8 per cent, had a cardiac illness, not one patient in the Kuwait group had any suggestion of heart disease. This is in line with the predisposition of the African in Africa to have a higher incidence of cardiomyopathy.

It is therefore apparent that Kuwait is no exception to the world-wide distribution of toxoplasmosis and may in fact be a comparatively heavy endemic area: Further epidemiological studies would be necessary to confirm this, but are justified because of the known importance of toxoplasma infection as a cause of abortion in cattle, and congenital disease in infants, leading to blindness and mental disease, comparatively common compared to the acquired disease reported above. An additional economic reason for further study would be to establish means of preventing the disease in domestic animals if large-scale stock raising plans are ever to be considered.

XI. Psychiatric Illness

"An asylum for the confinement of the insane was available in 1936, the main purpose of which was to keep the mentally sick from endangering the public and no regular treatment of inmates was really conducted." [1]

The effect upon a newcomer to an undeveloped or developing country of the acute and often violent manifestations of mental stress and disease, can be both startling and threatening, aware as he or she may be of the more insidious onset of the depressive disorders and the schizophrenics' behaviour in a mature society. Nevertheless despite the aggressive nature of many of these reactions, unlike the experience in a developed society, the ultimate prognosis can be good without resort to long continued drugs provided the conditions are recognised, handled expertly, and patients rehabilitated into the community, —often the most difficult operation. *In Kuwait the pattern* of mental disease still bears many of the features of a primitive community as suggested by the quotation above, taken from a Government publication. However, such has been the impact of modern psychiatric diagnosis, and in particular drug treatment, that by 1963 Kline [2] was able to say: "For Kuwaitis who come from the poorer classes, jobs are invented or overstaffed. For instance, there are ten guards for the grounds of the mental hospital who are paid $ 150 to $ 170 per month when in point of fact no guards are needed."

Before the advent of modern medicine and perhaps still among isolated communities, the mentally ill shared the same awful fate of neglect as their equally unfortunate physically ill brothers but with the added misfortune of continuing to "live" a life of confinement. It is worth quoting extensively from Kline [2], for nowhere else is the traditional method of treatment portrayed: "A neuropsychiatric patient (including epileptic) when he becomes ill is usually first taken to a mullah (teacher) at whose home he may stay for two or three days during which time the mullah reads to him from the Koran, blows on him in a prescribed ritualistic manner, or may write certain words on paper which are then wrapped in cloth and suspended on a string from the patient's neck or arms. The writing usually consists of magic symbols or words from the Koran and is done with a special ink of characteristic colour and smell, made by mixing prescribed desert herbs. If this fails the patient is often taken to a muttawe (native doctor) who may use the same type of treatment or may use kai, which consists of cauterization of the head, usually in the form of a cross from forehead to occiput and ear to ear or a circular line around the circumference. For non-psychiatric diseases other parts of the body may be cauterized and there are frequently specialists in one disease or another".

"The native word for mental disorder is sabab which literally means 'cause'. The term sabab is also used in reference to any other mysterious disease and 'cause' is believed due to 'underground devils'. In the past those cases of failure by the mullah and the muttawe were frequently chained underground in a cave or a closed room and on occasion attempts were made to drive out the devils by flogging. Only if treatment by the muttawe failed or when the family ran out of money was the patient taken to the mental hospital. There are still a number of quite famous muttawe in Kuwait City itself and about six of them are particularly prominent. However the city dwellers are beginning to lose faith in this method of treatment."

"Bedouins also recognise mental illness as such, and in the desert the muttawe in addition to or instead of kai may also use zar as a major treatment. This consists of a gathering of ten to twenty sick people, usually hysterics, at which the muttawe with some ten assistants acts as the master of ceremonies at a ritual 'drumming out of the devil'. It is up to the muttawe to decide what sort of demon is possessing the patient, because different rhythms and different sacrifices, e.g. a black hen, a brown and white lamb, et cetera are called for in different illnesses. The group, numbering thirty to forty persons including spectators, starts dancing in the early morning and may literally go on for two or three days until the patients drop exhausted. Fire and incense are used to heighten the effect, but curiously there is no report of drug usage for this purpose."

The asylum of 1936 was converted in 1945 to a hospital for mental patients but until the present director of the mental diseases hospital arrived in 1953, no specific treatment had been offered to the patients occupying the twenty-five bedded "hospital" which was still a place of maximum security with no proper hospital facilities or professional attendants. Gradually change and improvements were made and accommodation for females provided, hitherto non-existent.

In 1958 a new 200-bedded Hospital For Nervous Diseases, on the pavilion style, was opened and has proved of immense benefit to the people of Kuwait. Such was the pressure on its capacity that the old buildings continued in use for some years until it was possible to increase the number of beds to nearly 500, the present complement and almost always fully utilised.

The overall *incidence of mental disease* in this evolving society does not differ appreciably from that in Egypt or the western world, but the pattern certainly does. Communication between Egyptian psychiatrists and Badu patients was difficult in the earlier days; problems arose in distinguishing psychotic disease in these people in view of their manner of thinking and reacting and this was exemplified by the eventual recognition of a relatively high incidence of hebephrenic schizophrenia, characterised by the assumption of curious mannerisms. On the other hand they have rarely seen paranoia in desert dwellers, in contra-distinction to the picture in immigrants into the country. The following is a list of the most frequently experienced diagnoses, including organic brain and nervous system disease. They are in order of prevalence among admissions to the Hospital For Nervous Diseases.

The predominance of Kuwaitis is outstanding despite the relative equality of Kuwaitis and non-Kuwaitis in the country at the time.

Darwish [3] has reviewed his experience among seventy-nine schizophrenic patients seen over a period of six months in Kuwait. There was a significantly higher incidence of delusions among the urban population; the role of social factors, in particular the disruption of family and community life, has contributed to this, as has been suggested for other pastoral and rural people who find themselves beyond reach of the security and mental comfort of their familiar environment.

Table 48. *Diagnostic distribution of most common conditions in 1960*

Schizophrenia	156	Addiction to morphine,	
Manic depressive		opium, heroin	7
psychosis	110	Senile dementia	5
Alcohol addiction	29	Mental deficiency	
Hysteria	23	(non-educable)	5
Anxiety state	16	General paralysis	4
Paranoid state	15	Parkinsonism	3
Epilepsy	8	Other non-specified	21

Table 49. *Nationality of admissions to nervous diseases hospital*

Kuwaiti	231	Iraqi	30	Iranian	24
Jordanian	34	Lebanese	16	Indian	8
U.A.R.	11	Palestinian	5	Mahri	10
Saudi	24	Omani	19		

The advent of the psychopharmaceutical *drugs* has, as elsewhere, revolutionised the handling of both the acute maniacal phenomena and the longer term rehabilitation, although impressions are that drugs do not require to be given for long periods. With the inevitable increase in urban social problems following the rapid population expansion and consequent pressures, a good long-term prognosis may require prolonged drug therapy. There are already indications of this change.

An observer who first visits the Middle East may be startled to see the freedom with which young men hold hands in public and exhibit obvious affection even to the extent of kissing. On the other hand, overt affection between the opposite sexes in an orthodox Mohammedan country is taboo and could lead to punishment by both the woman's family and the State. The high incidence of *homosexuality* superficially suggested by hand holding is confirmed by the neurotic effects which such practice leads to. This habit is not, as traditionally thought, confined to the Badu but has become part of the culture of urban dwellers. As in other countries where heterosexual relations are restricted by rigid religious, moral and social intolerance coupled with the relative poverty of the young men, sexual satisfaction is achieved by masturbation, sodomy and bestiality. That such habits are not restricted to Arabia can be confirmed by those who have experience of treating venereal disease in other countries: The price of a bride may be beyond the means of many men of the lower social strata and they turn to the only means of satisfaction left to them. This is another example of the tragic consequences of population explosion following public health improvements which are not supplemented by equivalent evolution of the traditional customs and prejudices; these were originally designed to protect a community continually subject to disruption and destruction. Those concerned with psychiatric care in Kuwait have noted an interesting, though not unique, dissimilarity between the neuroses of Badu and Kuwaiti men practising homosexuality; the former seldom exhibit paranoid features whereas the latter frequently do. It has been suggested that the absence of a rapidly evolving western way of life and education in the Badu protects him from the distressing paranoid features, frequently seen among urban Kuwaiti and other westernised Arabs. The clandestine pattern of homosexuality among the latter groups is accentuated by the loosening effects of forbidden alcohol, which are frequently accompanied by homosexual relationships. The desert, prior to prohibition in 1964, bore witness to many alcoholic excesses for it has been said that the empty cans of beer litter the ground sufficiently to hide the sand! Elsewhere we have remarked that recently young men have been taking wives from outside Kuwait, in particular from western Iraq, and one can hope that this new practice will lead to a decline in the old.

Another aspect of the rigid social custom of female segregation is the sometimes alarming frustration which lurks behind the Burqa, the coarse black silk mask worn by Badu women in Kuwait. But frustration is also felt by many well-to-do women living a life of boredom while their men conduct the thriving business life in the city. Few outlets have hitherto been available to Kuwaiti women but the influence of the education of the young woman is now creating strong pressures to open up avenues of activity previously reserved for men. These very pressures may result in a change in the present ratio of two Kuwaiti men to one woman admitted to the Nervous Diseases Hospital. Heretofore the woman has been protected from the rough and tumble of competitive urban life and her boredom has been reflected in the many *psychosomatic complaints* from which she suffers. Perhaps the commonest of these are related to the musculo-skeletal system, which is often neglected and appears to be making a vicarious appeal for greater activity. A secondary factor has been that such women often become excessively fat, a feature appreciated by the Arab husband but conducive to joint pains and muscular weakness. That stress of the nature we have described can perhaps also act in a western mode is illustrated in the accompanying photograph of a Badu woman with advanced rheumatoid arthritis; we have no evidence that she had in fact undergone any unusual or significant stress: Nevertheless it was felt she was also suffering a depressive illness as might be guessed from her appearance (Fig. 51).

The specific *effect of industrialisation* upon the young Kuwaiti has been observed in the petroleum industry (Fig. 52). The 20—40 age group have found themselves exposed to stresses during the past ten years for which they had been quite unprepared; over this period rapid Arabisation, technical training, and the assumption of administrative and managerial responsibility have had an inevitable effect; many casualties have occurred. Gradually it became apparent that careful selection for initial employment with its potential for advancement and for re-appraisal of current employees before considering promotion was mandatory if those selected were to be expected to carry the burdens of the future. Perhaps one of the most disturbing features has been the lack of criteria by which to measure the adequacy of candidates, emphasised when selecting young men for advanced technical training overseas. It is to be hoped that lessons have been learned, for Kuwait cannot afford to dissipate its brain-power even if it can afford to lose some of its investments. One of the most encouraging features of the period has been the close understanding and cooperation which developed between the staff of the Government psychiatric units and the doctors in the oil company for the latter is the life-blood of Kuwait and the future stability of the industry is going to depend upon the young Kuwaitis and other Arabs of today who are capable of adapting to meet the unforeseen.

XII. The Haemoglobinopathies

1. First Recognition in Kuwait

Elsewhere we have remarked upon the interesting analogy of Kuwait to a "genetic cockpit" and in no group of conditions is this better illustrated than the disorders of haemoglobin synthesis and enzyme activity in the red blood cell. Their significance in Kuwait was first recognised in 1951 by Dr John Walters, but laboratory facilities at the time were not sufficiently developed and it was another ten years before any reliable evidence became available to reveal the fascinating mixture of genotypes which are today walking the streets of Kuwait. These are the successors to the pastoral nomads who roamed the desert wastes for thousands of years before the new wealth from oil ended for ever their traditional pattern of life and created another Mecca across the great Najd desert, to draw Arabs from every part of the Middle East and beyond. Yet the patterns of haemoglobin disorders which are seen today may appear pallid beside those of tomorrow when the puny infants who are now permitted to survive meet together, marry and reproduce their kind. It is worth pointing out that this success is not achieved without cost, for the medical care and continued attention to these people is considerably greater than that required for the average population.

The pattern of the *world distribution* of these conditions has occupied the debating time of students of epidemiology and anthropology for a number of years, particularly as it approximates remarkably closely to that of the distribution of malaria; a look at the map demonstrates that Kuwait stands close to endemic malarious areas and may once have been a focus itself. More likely however that the mixed pattern of haemoglobinopathies stems from the remarkable linkage of populations which has taken place over some hundreds of years. That these mutations in some way afforded a protection from malaria to a population by a natural resistance, is now well accepted but the arguments which have led up to this agreement should be studied in definitive texts, such as Lehmann and Hunstman [1].

The few cases of sickling disease which were identified twenty years ago, were, unfortunately, not followed up and were soon lost to view once interest declined because of lack of refined laboratory facilities: These were only then beginning to be used in more sophisticated laboratories elsewhere. By 1961 the laboratory services of the Southwell Hospital had introduced accurate methods of haemoglobin electrophoresis, identification of foetal haemoglobin and red-cell enzyme analysis and within a short time a high degree of skill was developed.

It had been apparent for some while that a number of patients previously untreated and presenting with anaemia in association with other conditions, notably gastro-intestinal and respiratory disease, showed evidence of haemolytic anaemia. These were now re-investigated and very rapidly the picture of sickle-cell disease, sickle-cell trait, thalassaemia major and minor and Glucose-6-phosphate dehydrogenase deficiency emerged. This early review of the problem in Kuwait drew much stimulus from the work of Dr A. P. Gelpi [2] in the Eastern Province of Saudi-Arabia and Professor Lehmann at Cambridge [3], England. It was realised however that one great difference existed between Kuwait and the surrounding Arab countries, for the latter generally were plagued by falciparum malaria, whereas Kuwait was entirely free from endemic infection. This required explaining in view of the prevailing hypothesis that these haemoglobinopathies provided a selective advantage for populations in malarial zones who underwent a gradual change in genetic composition through a number of generations until a stable genetic pattern emerged in a geographically isolated or segregated population, the phenomenon of "genetic drift". Kuwait however has exhibited a completely contrasting situation where, for many years, even before the advent of oil, a physical drift of population from the desert towards the entrepôt port of Kuwait has been in operation, and this drift has come from many different directions by land and sea to create the polyglot population which is the urban people of Kuwait today. The varied source of these people is the clue to the rather remarkable prevalence of the haemoglobinopathies recently uncovered, and is only part of the larger picture, albeit patchy, of such loci in the Near and Middle East. Almost yearly new identifications are being made which is not entirely unexpected in view of the area's strategic, geographic and religious significance, lying in close relation to the three continents of the old world and along the pilgrimage routes of the three great monotheisms, Judaism, Christianity and Islam [4]. While Kuwait is malaria-free today, that has not necessarily always been the case, for we know that this coast, and Failaka Island particularly, once had good sweet water and was well treed, two pre-requisites for mosquito breeding. Anopheles pulcherrimus, which has a tolerance to an arid environment and will breed in water with a salinity up to 6,000 parts of sodium chloride per million, is still occasionally found but has never been known to transmit malaria in the neighbouring countries. On the other hand, in similar saline marshy areas to those surrounding the northern shores of Kuwait Bay, elsewhere along the coast of the Persian Gulf, A. multicolor, a probable malaria vector in Iran, could in the past have been responsible for endemic infection in Kuwait.

2. Sickle-Cell Anaemia

In all probability the sickling phenomenon occurred in the pre-semitic population of Arabia which moved east to India and south to Africa leaving, in neolithic times, pockets of people with this characteristic haemoglobin behind. Sickling is due to the deoxygenation of an abnormal form of haemoglobin, HbS, present in varying amounts in the red cells of affected individuals and replacing the normal adult type of haemoglobin, HbA, in amounts according to whether the individual carries the heterozygous state (sickling trait) or the homozygous state (sickling disease). In a deficient oxygen state the HbS becomes changed to crystalline-like structures or tactoids which cause the altered shape of the red cell due to increased rigidity. With this altered shape arises an increased tendency of the cell walls to break and release the contained haemoglobin (haemolysis) and to clump together (thrombosis), which in turn leads to the symptoms and clinical phenomena seen most usually in sickling disease, but occasionally in carriers with sickling trait.

The impact of sickling disease in Kuwait upon the interested observer must be relative, for it cannot be compared to that experienced for instance, in West Africa, where the homozygous state will create a lasting impression by its frequency and multiplicity of accompanying severe disorders. Nevertheless, in Kuwait, the manifestations are relatively frequent as illustrated in the two series of cases which have been studied [4, 5].

The remarkable degree of consanguinity which occurs in Kuwait is certainly reflected in these two series, occurring as often among the Palestinian patients as among the Kuwaitis.

The age range was from $1^{1}/_{2}$ to 20 years and the sex ratio predominantly female (29 : 11). Hyperteleorism, negroid and mongoloid features were observed among the two groups and the following complaints were recorded:

Inability to walk	Swelling of joints
General weakness	Muscle pains
Limb pains	Acute abdominal pain
Red urine	Acute renal angle pain
Backache	

These would occur either singly or associated in any combination. Splenomegaly was frequent, and bone changes at x-ray gave pictures typical of sickling disease, but to the uninitiated so easily confused with osteomyelitis. Anaemia was marked in many patients at first presentation and foetal haemoglobin (HbF) levels varied from nil to 30%: One patient aged two days had 73% HbF. An interesting patient, a male Kuwaiti aged 30, had reported sick complaining that he and his wife were unable to have any children. He was found to have azospermia and blood-stained seminal fluid, the red cells of which sickled, thereby leading to the diagnosis of his trait. This is thought to have been a unique observation.

We have remarked earlier that there was a high degree of consanguinity in these patients' family histories: This often leads to multiple defects. In the series of Ffrench and Shalhoub [5], one patient had associated gross deficiency of Glucose-6-phosphate dehydrogenase: The association of this deficiency with thalassaemia was also noted. Accompanying congenital heart disease occurred in a number of cases in both series.

The causes of the severe symptoms are thrombotic crises and once the possibility of sickling is kept in mind, the diagnosis is not difficult. It is in the less clear cut symptoms associated with the trait that the diagnosis has often to be made by exclusion.

The well-known association of sickling with salmonella septicaemia was seen in two cases: Salmonellosis is relatively frequent in Kuwait and the hypothesis that these organisms find their way from the gut to the blood stream as a result of damage to the gastro-intestinal mucosa from areas of sickling thrombosis would seem to be supported.

Treatment of the crises is urgent and difficult for they can be rapidly fatal. Ffrench and Shalhoub began the use of the alkali and magnesium sulphate regime [5] on the suggestion of Lehmann [3]. 1—2 ml of intravenous 50% magnesium sulphate was at once seen to be of immense value in reducing the severe pain and averting shock; the vaso-dilator effect assists in dislodging clumps of sickle-cells from the small vessels. 2—3 Gm. of sodium bicarbonate in divided doses during the day raises the alkali reserve and, in the absence of gross deoxygenation, will maintain the blood free of sickling. Random sampling of the urine was carried out to check on the continuation of alkali ingestion. Analogous to the use of steroids in other diseases at the time of intercurrent infection or other stress, the dose of bicarbonate should be increased sufficiently to maintain the urine alkaline. In this manner several of our cases were safely brought through later surgical treatment.

The prognosis of these patients will depend upon the quality of their follow-up care: It is expected that adult homozygous disease will be seen in the future in Kuwait. This will pose problems in marriage counselling and in the management of pregnancy but with the standard of medical care to be expected, these difficulties are not insuperable. It can therefore be foreseen that, due to genetic effects, sickling trait and disease will continue to increase in the foreseeable future due to both the survival and fertility of people with sickle-cell disease.

There is a strong case to be made for nation-wide identification of haemoglobinopathies because, armed with this knowledge, advice can be given and implemented in such a relatively well identified population.

3. Thalassaemia

The amount of sickling disease in Kuwait is not to be compared with that of thalassaemia and it is this latter condition which we believe may have a profound effect upon the ability of a relatively heavily affected population, both autochthonous and immigrant Arab, to withstand the pace of change to a modern competitive industrial state.

There is now a growing opinion that much of the infectious and non-infectious disease in infants and children occurs in patients already the subjects of imperceptible debilitating nutritional, protozeal and genetic disease: we believe that thalassaemia illustrates the effects of the last of these very clearly and in a not insignificant portion of the population.

In thalassaemia there is not a replacement of normal haemoglobin (HbA) by an abnormal one but rather a suppression of its synthesis. This happens because one of the polypeptide chains suffers a defect probably due to a failure of genetic communication. HbA consists of alpha (a) and beta (β) polypeptide chains: It is more frequently a defect of the latter which is found in the classical anaemia of the Mediterranean basin. Thus the naming of the disease from the Greek word for the sea, which, as far as the ancient Greeks were concerned, consisted only of the Mediterranean. Nevertheless it is distributed far more widely. Defect of the alpha-chain has also been found, but no loci in the Near and Middle East have yet been identified and it does not play any part in the haemolytic disease of Kuwait. Lehmann says that Beta-thalassaemia is the most widespread abnormality of haemoglobin production and that virtually every population shows some examples "Although it is notably rare in Africa" [3].

As the basic abnormality is in the Beta chain polypeptide, the other two components of the red cell at birth, the foetal haemoglobin HbF, and HbA2 are present in relatively higher concentration. HbF can, in infancy, carry out the functions of normal haemoglobin efficiently and protect the infant for several months; after this time it diminishes. But when there is no concomitant production of normal adult haemoglobin (HbA), symptoms may then "burst" upon the child, depending on whether he or she is homozygous for the gene deficiency (β-thalassaemia major) or heterozygous (β-thalassaemia minor). Whereas the former is very severe and, until recently, almost fatal within a few years, the latter is usually only associated with a mild anaemia unless complicated by intercurrent disease.

Foetal haemoglobin (HbF) comprises 45—90% of the haemoglobin of the newborn but normally decreases rapidly during the first year: It is rarely demonstrable after the age of thirty months. However, in some people the mechanism for foetal haemoglobin formation may never be totally lost and thus a figure of 2% can be regarded as the upper limit of normal. Large quantities, up to 15%, are synthesised in patients with homozygous sickle-cell disease, and in severe β-thalassaemia major it may represent almost the entire functioning haemoglobin, albeit a relatively inefficient one. It may also be found in other forms of hereditary anaemia in small amounts, in some patients with leukaemia, nutritional anaemia and more rarely it may be a familial or hereditary disorder.

Clinically the homozygous state, β-thalassaemia major, has been until recently almost invariably a fatal disease: it is accompanied by severe debility and disability from a few months of age when the formation of physiologically active HbA fails to take place at the time the HbF has begun to diminish.

With the failure to develop normal HbA, the red blood cells not only have a reduced haemoglobin concentration, which may also occur in iron deficiency anaemia and then to a much greater degree, but the cells themselves are morphologically abnormal in shape, being thinner and in consequence more fragile, having a shortened life span less than the normal hundred and twenty days. These thin cells are characteristic of β-thalassaemia major and are known as target cells; when seen in a blood smear, they are almost diagnostic. This is certainly so in an endemic area. Because the cells are fragile, they disrupt easily, accounting for their short life. A continuous wastage of functional haemoglobin into the plasma occurs, creating a state of chronic haemolytic anaemia. This in turn causes a compensatory overactivity of the bone marrow resulting in enormous expansion of the sites of bone marrow activity, normally restricted to the long bones and ribs, but now filling the medullary cavities of the cranial and facial bones (Fig. 53). The patients develop a characteristic facies with wide separation of the incisor teeth, a phenomenon analogous to the characteristic frontal bossing of the skull in homozygous sickle-cell disease, for in thalassaemia it is maxillary hypertrophy which predominates. When the balance between red cell destruction and bone marrow compensation cannot be maintained, increasing anaemia and death from intercurrent infection takes place. Together with the bony changes there is very often a gross enlargement of the spleen and to a lesser extent of the liver, in both of which organs compensatory red blood cell production arises, but is accompanied in the spleen by an over-activity of the destruction of red blood cells and other cellular components of the blood; the phenomenon is known as hypersplenism and contributes greatly to the parlous state of the child's ability to survive.

The treatment consisting of attempts to reduce the inexorable progress of the disease is beyond the scope of this book: Enough to say however, that recently in Kuwait it has been possible, with the excellent facilities available, to study the care of these unfortunate homozygous children and to provide treatment for the complications as they arise, in particular the hypersplenism and the iron overload resulting from red blood cell destruction [7].

The heterozygous β-thalassaemia minor is worldwide in distribution. While the clinical state of a wellfed and otherwise healthy carrier is almost as good as an individual with normal haemoglobin, the former does have a mild anaemia which, in the presence of infection or parasite load, can cause appreciable effects on work output. The haemolytic anaemia is present though mild and the changes in bones and spleen are much less evident. When Lehmann and Huntsman (1966) [1] have said "They live on the edge of a volcano and it is the duty of the physician to listen out for the early rumblings that may spell early disaster", they are referring to the precipitous anaemia that may occur in these people on whom additional load is placed by intercurrent disease or physiological strain such as pregnancy.

The diagnosis of the haemoglobin abnormalities has been a fascinating epoch in man's groping towards perfection. Details can now be found in standard texts and should be referred to. It requires us only to say that the methods demand a training, application and mechanical aids which may often be beyond the capacity of small understaffed hospitals and clinics, but which nevertheless should be a high priority in those parts of the world now identified as endemic or into which considerable immigration has occurred from endemic areas.

A knowledge and understanding of the basic differences and particularly the clinical significance of sickling disease and thalassaemia is mandatory for anyone claiming to practice successfully in those countries of the world with a proportion of the inhabitants affected. While the homozygous states make an immediate impact upon anyone unfamiliar with their picture, the heterozygous states are often accompanied by a delicate balance of health; this is particularly the case with β-thalassaemia minor. They explain much of the apparent weariness and indolence of tropical people who have to get along with a reduced amount of physiologically active haemoglobin, often on top of a parasitic load, which by itself would be sufficient to curb the energy of a full-blooded man, woman or child. In our opinion, this aspect has been insufficiently appreciated, particularly when industrialisation, with its physical and mental demands, requires efficient and often sustained output.

4. Glucose-6-Phosphate Dehydrogenase Deficiency of the Red Blood Cells

This rather terrifying title describes the first recognisable and the most widespread of the enzyme deficiencies of the red blood cell to which man is subject and as a result of which he may suffer acute effects when exposed to certain trigger mechanisms. Wyngaarden (1966) [8] described it as follows, "The G6PD deficiency states comprise perhaps as many as a dozen sub-varieties, in which the enzyme differs chemically or physically from the normal. There is every reason to believe that some of the G6PD or pseudo-choline-esterase variants fall in this class—i. e. simple amino-acid replacement in an enzyme". Kuwait contains the variants affecting both the Mediterranean peoples and that of African populations.

This enzyme catalyses the first rate-controlling step in the hexose-monophosphate shunt metabolic pathway of the red blood cells, and also in cells of other tissues. Its deficiency is related to the reduced glutathione content of the cell and in the case of the red blood cell, to the stability of haemoglobin, and leads to failure of formation of adenosine triphosphate, a co-enzyme essential for maintaining the ionic composition of the cells and their cellular integrity. This is followed by haemolysis and leakage of haemoglobin into the plasma, recognisable as haemolytic anaemia. The hexose-monophosphate pathway is the oxidative one, though comprising only 10% of the total energy production: It is essential for maintaining haemoglobin and glutathione in their reduced state and so guaranteeing the oxygen carrying capacity, the glutathione particularly being of importance to the survival of the cell.

The deficiency is inherited as a sex-linked character, the responsible gene being located on the X chromosome; thus only males exhibit the full clinical phenomenon. Of the specific tests to detect deficiency not all are capable of identifying the heterozygous state in the females and these themselves may differ in their capacity to react to the tests, because of the mosaicism which is not unusual in this condition in females, some cells utilising the normal and some the abnormal chromosome.

People with this enzyme deficiency do not usually develop haemolytic anaemia unless exposed to certain trigger agents. The haemolysis, unlike that of the haemoglobin variants, is self-limiting because only the older cells are broken down, the younger cells retaining a relative sufficiency of the enzyme. These trigger mechanisms have been identified as certain drugs, the broad bean, Vicea fava, grown and eaten throughout the prevalent areas and noted by us to give rise to haemolytic crises in Kuwait in people eating tinned beans from Ethiopia out of season, certain virus infections such as the Eaton agent of atypical pneumonia, arthropod-borne viruses and more recently the virus causing infectious hepatitis. The possibility that exposure to industrial compounds both during manufacture and as pollutants may trigger a crisis is a real one in an affected population, for those compounds containing amine groups are potentially dangerous.

Several observers have identified G6PD deficiency as a major cause of otherwise unexplained severe neonatal jaundice and it is very important to exclude this possibility in Kuwait. Vitamin K or its analogues should not be given to the pregnant mother or the child as these drugs are known trigger agents: This stresses the importance of obtaining a family history of neonatal jaundice, favism (broad bean anaemia) or drug induced haemolysis which can alert the physician to the chance of disaster. Favism is of interest historically because for many years an acute anaemia had been noted in Iraq during the spring, finally identified as a form of favism due to the inhalation of the pollen of Vicea fava, the broad bean. This vegetable is imported and consumed in quantity in Kuwait.

The earliest survey for G6PD deficiency in Kuwait was reported by Shalhoub and Ffrench [9] in a pilot sample and was sufficiently impressive to make them include the tests in screening procedures being carried out for anaemia. What is also impressive is the combination of this deficiency with other congenital and genetic defects in an individual, giving him or her an almost impossible chance of survival but which a proportion manage to do! In 1966 further work in Kuwait [10] confirmed the importance of this condition by establishing a prevalence of nearly twenty per cent in native-born Kuwaitis and just over twenty per cent in the overall population, comparing well with similar percentages found in neighbouring Saudi Arabia by Gelpi (1963) [2].

Once again the awareness of the condition is all-important for a correct early diagnosis. Retrospective assessment in some areas has led to the review of previously accepted concepts of blackwater fever in apparently immune Africans in holoendemic areas. Gilles (1964) [11] had subsequently found the enzyme deficiency in a number of cases where the phenomenon of blackwater fever in an immune African seemed to demand an explanation other than that of hyperinfection with *Plasmodium falciparum*. Such a retrospective approach is invaluable in clearing up some previously puzzling questions in the light of current genetic research.

XIII. Heat Illness and Desert Survival

Eskimos live with reasonable comfort in sub-zero centigrade temperatures inside the Arctic Circle and Badu in the Arabian and North African deserts have survived for many hundreds of years living at the other extremes of temperature in excess of 50° Centigrade. Their secret is threefold; adaptability, controlled expenditure of energy and the development of a satisfactory micro-climate. Man, like other animals and plants but more successfully, has learnt to adapt as a study of the various arctic and desert inhabitants has shown, and they reveal many analogies. In the Arabian desert the above ground flora are xerophytic with long penetrating roots designed to reach the moisture deep under the sand while the leaves are xerophyllic, very tough and presenting the smallest possible surface to the abrasive winds. Many of them have at the same time developed thorns to discourage scavengers and enable them to survive. The animals on the other hand, are either burrowing or have developed remarkable powers for the conservation of fluid and electrolytes, and the loss of heat. The burrowing animals have adapted by becoming virtually nocturnal, returning in the day to their underground home which is maintained at a remarkably constant temperature both in winter and summer. The few animals remaining on the surface either have favourable body weight to surface area ratios, together with specially adapted heat losing organs, in particular the very large vascular ears or pinnae, or as in the domestic animals, a woolly coat which insulates against the heat as well as the cold (Fig. 55). The only exceptions to obvious physical adaptations are the camel and the ass, but these have some curious mechanisms to offset their potential liability to the effects of the terrible desert heat.

The *Badu* would seem to have been living for a much shorter time in the desert than the indigenous animals and has not yet had time to develop overt superficial changes: When naked he looks much as other men although often bearing a remarkable resemblance to some of the xerophyllic plants! However, in contrast, his human powers of reasoning have enabled him to learn from his experience; at the same time he has retained the remarkable powers of instinct which enable him to be at home in what, to a stranger, appears a land without colour, character, landmarks, or sanctuary. This capacity is exemplified by the Murra tribe and members of the Sulabba community. But the Badu has learnt to know his desert, to respect its dangers and to appreciate that if he is to survive he must conserve his energies because by doing so he will reduce heat gain and water loss. There is no inherent modification in the anatomical or physiological make-up of the Arab desert dweller analogous to the steatopagy of the Kalahari bushmen whose association with desert life derives from far more primitive times. The Arab has adjusted his pattern of living and in particular his religious and social customs to allow him to survive in what, before his recent urbanization, he used to regard as reasonable comfort. The danger today lies in the fact that the Badu is fast becoming detribalised and a town dweller: He will lose perhaps more than he will gain for there is no going back in this present civilisation, although it is the very desert itself that emphasises the impermanence of successive civilisations by the evidence which lies under its sands. Yet still the urbanized Badu appears to exhibit an urge to return to desert life, though much later in the evolutionary time scale than the desert aborigine of Australia who periodically goes "walk-about".

We have said that the Badu conserves his physical effort during the time of the sun's greatest heat: He does this by remaining quietly in natural or artificial shade and within his own little climate inside his loose and flowing robes. By these means he achieves a nice balance between heat gain and heat loss and retains his energies for the cool hours of the day. This kind of mental and physiological control over his activities is only possible in a simple pastoral culture where the minimum of husbandry is required. Man's good sense under these conditions is mirrored by the remarkable design of his desert house, his tent. This is made of woven goat hair and sheep wool of considerable thickness and weight. It is always pitched with its long axis facing the wind, either the prevailing north-west (shamal) or the humid south-east (kaus). Thus it is that almost always in Kuwait and among neighbouring desert tribes, the long axes of tents will all be in the same direction, on a north-east or south-west axis. The Badu takes great care over this and sees that the side of the tent facing the wind is closed, and the sheltered side open. When in fact there is little wind, all the sides will be rolled up and such breeze as there is will waft through the tent: At the same time, a thick carpet may be raised inside the roof to act as a double skin. It is quite surprising, in the heat of the day and provided there is the minimum of physical effort, how relatively comfortable such a "house" can feel. There is also an elaborate system of customs or manners which, to an outsider, might seem ridiculous but when examined are designed to achieve the greatest comfort for the owner, his family and such guests as there might be. By such simple personal and household discipline the Badu of the desert have been able to survive what, to anyone unwilling to adopt his customs, would be certain death in the height of summer.

A surprising number of otherwise intelligent people have left themselves ignorant of the relatively simple principles of body heat conservation in a cold environment and heat dissipation in a hot environment. With many thousands now climbing mountains, sailing on seas and travelling to foreign parts, the cost alone of rescue and resuscitation has risen to serious levels. When leisure is now becoming an integral part of everyone's experience, it also becomes an individual responsibility to keep away from trouble or know how to deal with it.

We can learn a lot from the Badu. For instance, he shears his sheep a sufficient time before the hot weather to allow the woolly coat to reform to provide adequate insulation against the hot sun (Fig. 56). There is a temperature gradient of as much as 45° C between the surface of the skin and the wool: Beneath this coat the animal will live reasonably comfortably in a microclimate, the evaporation of sweat taking place at the skin surface. The woolly coat absorbs radiation heat which is in turn convected away by natural air currents or those generated by the movements of the animal: In the process the loss of further heat by sweat evaporation is facilitated. If this very efficient system broke down heat gain would exceed heat loss, and eventually the metabolic production of water, sometimes the only source, would be insufficient to supply the sweat: Heat stroke would inevitably follow.

The example set by some unitiated people when exposed to a hot, dry climate is unfortunate: The lack of sufficient clothing over the limbs and body is equivalent to a slow grilling from the radiant heat waves of the sun: To achieve any sort of heat balance requires an enormous throughput of water, sodium, potassium and other salts. The high rate of metabolism in turn burns up the carbohydrate, fat and protein stores and wears out the cardio-vascular system by the continuously high cardiac output required to maintain the enormous fluid exchange, particularly in the skin.

Man is able to survive and even be reasonably productive in the Arctic and the high desert temperatures of Arabia by a combination of internal metabolic compromise and external artefactual insulation during the hours of greatest exposure to cold, heat or humidity.

The human body has a normal temperature of 36.8° C which is maintained against the loss of heat to the surrounding air by the combustion of food in the body. This combustion also provides the energy for all activity, mental and physical. Mental activity uses comparatively small amounts of energy in contrast to hard physical work. In the process the body generates heat, which has to be dissipated to enable it to maintain its normal temperature. Excess of production of heat over dissipation results in a rise of body temperature, and if it goes too far, ends in heat stroke.

Heat gain can also arise by direct radiation from the sun reflection from surfaces such as rock, sand, water and snow; by conduction from hot surfaces, in particular sand and metal, and finally by convection from hot air.

In the same way *heat can be lost* very quickly to such surfaces and the air. The only means of dissipating heat under these circumstances is by evaporation of sweat which requires a saturation deficit of the surrounding air together with sufficient air movement. However, if the air temperature is significantly above 36° C, then the heat effect predominates, should sweat production and evaporation be unable to keep pace with it.

The measure of the balance between heat gain and loss, felt subjectively by the sensation of comfort, is the basis for the "corrected effective temperature" (CET) scale, which requires three measurements, wet-bulb and radiation temperatures and air speed. Obviously in the context we are discussing, very hot desert conditions, there is no comfort zone, and the recognised limits for heavy work are exceeded. For practical purposes production of sweat, clothing, an intelligent appreciation of the proper expenditure of energy and care of the skin are the main criteria by which survival in a hot dry and hot humid conditions can be achieved, although other factors such as age, sex, body build, physical fitness, the type of food, drugs, alcohol and recent illness must all be accounted.

The eccrine sweat glands, which occupy all the body surfaces except the arm-pits and the groins, are the thermo-regulatory glands and this is their sole function. Daily losses of water in sweat may be as high as 8 kg and of salt 20 Gm. The only replacement in the human is by drinking and preferably the addition of salt to the water. The tremendous cardiac output required to circulate the blood through the skin to achieve this degree of water and salt loss puts a heavy load on the cardio-vascular system. Therefore heart disease in this type of climate exposes an individual to serious risk.

Early in the development of modern Kuwait it became obvious that an understanding of survival at high environmental temperatures was mandatory if exploration for oil and further development of production and sea transport was to take place. This realisation came at a time when physiological problems of heat were being studied in some depth quite literally, by the British Royal Navy in regard to submarine crews, the British Army in relation to desert warfare, by industry in deep mines and by the steel industry. But there was little coordination of ideas or effort. For the Kuwait Oil Company the problems were twofold in the main, hot humid atmosphere affecting port workers and tanker crews on the one hand and very hot but dry desert conditions during five summer months. A concerted approach by experiment and education enlisting the help of the Liverpool School of Tropical Medicine produced a very significant improvement both in the morbidity from heat illness in the company workers, many of them expatriates, and among tanker ships' crews.

Unfortunately it was not then possible to study the indigenous Arab and particularly the Badu in Kuwait. As the years passed it became evident that the latter was losing his unique desert adaptation once he had ceased to live his nomadic life. Many are the tales of the Badu's extraordinary ability to conserve water and to a lesser extent salt but there has been little scientific evidence to support them. A recent investigation in Jordan [1] to test the truth of these tales was able to be as dogmatic as this: "Few peoples are surrounded by greater mystique than the Badu of Arabia. Their courage and bravery, their customs, and above all their incredible ability to

survive in the inhospitable conditions of the desert have been described and discussed exhaustively in books and travellers' tales. Burchardt, Palgrave and Doughty wrote about them in the nineteenth century; Lawrence and Thesiger have done so more recently. Their activities during the world wars enhanced their reputation for endurance in the desert, *yet there has never been scientific proof of this.* In 1966 it was possible to write: 'Apart from Israel, no reports have been found which describe specific studies on the indigenous people of South-West Asia and their response to heat.' [2] Recently an attempt has been made to repair this deficiency and the authors learnt more than scientific facts." One must understand however that the introduction of modern medicine and science into Arabia is of very recent origin and that it has neither been possible nor indeed wise to approach a proud and in many ways intolerant people with the object of conducting scientific research which would not be seen to have an immediate beneficial effect upon them. For example the with-drawing of a blood sample can be interpreted only too readily as an attempt to emasculate the donor. The scientific report quoted above did in fact reveal that "The Badu Arab subjects did not show a greater ability to conserve fluid by renal mechanisms than the non-Badu Arabs and Europeans." What in fact the investigation did show was that the Badus' understanding of the word "fluids" extended a considerable way beyond water, for despite agreeing to restrict water they did in fact consume tea, coffee, soup and camel's milk which contributed to their robust results in the water deprivation tests. This type of error probably accounts for the misunderstandings of many actual observers who have been quite dogmatic in their statements that the Badu can exist on as little as "half a mugful of water per day". Such a one is Wilfred Thesiger, but he tells us in his books that the Badu will frequently milk his camels and slake his thirst. Having had personal experience of this delectable fluid we can understand the Badu's preference for this over brackish water.

Replacement of water is never complete at the time of loss and has to be restored later: Thus a gradual water debt may build up as the period of loss is extended. There is therefore a limit to the period during which adequate heat loss can be sustained, depending upon water and salt intake, acclimatisation which involves the progressive dilution of sweat and restriction of salt loss by aldosterone control, the cardio-vascular reserve and finally voluntary control over effort (heat production). There are other regulatory mechanisms, such as nervous control via the sympathetic nervous system through thermoreceptors and the hypo-thalamus, endocrine via the anterior and posterior pituitary and the adrenals, heat exchange through the lungs and ingestion of food and drink, but these are beyond serious practical consideration.

A brief word is now in order to describe the various clinical conditions provoked by heat stress:

Water Depletion: the effects of this becomes obvious when two to three per cent of the body weight is lost. Beyond ten per cent the mind becomes clouded, there is inability to drink because of the dry mouth and throat and collapse followed by heat-stroke is not far away; it is at this point that an individual is unable to help himself and will die unless aid is given. Much can be done to prevent this situation by frequent small drinks of cool (15—20° C) water and not waiting for a water debt to build up: Under conditions of stress up to fifteen litres a day may be required which illustrates the degree of fluid turn-over, for this is three times the blood-volume. When the clinical condition arises, then oral or intravenous replacement is urgently needed, preferably of 0.45% or 0.9% sodium chloride solution, depending upon the salt deficit which may have to be assessed by the circumstances in which the victim was found. Salt deficiency is not likely to be a major problem in those stranded in the desert *providing they have not carried out hard manual labour or walked great distances,* or in those who have been lost at sea for any length of time, with the same qualifications.

Thirst Fever: One group of individuals who are very sensitive to relatively minor degrees of water deprivation are infants, and a characteristic pattern of their reaction to this in Kuwait has been described by Shaker [3], confirming the work of others, particularly in Greece [4]. The result of an absolute or relative absence of water is inadequate heat elimination; the Kuwait series of seventeen cases were aged from three to ten months, were all otherwise quite healthy and were receiving their correct estimated daily requirement. The fever was explained by a state of relative water insufficiency initiated by the hot humid weather and other factors such as high caloric or protein intake, the possibility of subclinical rickets with excessive sweating and the restriction of fluid intake overnight after a last evening feed—thus the temperature characteristically is highest in the morning.

Shaker [3] noted that Kuwaiti infants very seldom suffered from thirst fever, and explained this by the fact that Kuwaiti mothers are in the habit of giving extra fluids to their infants from birth.

Salt Depletion: This will occur when water replacement is not accompanied by the salt lost in sweat. Characteristically it is associated by fatigue, headache, anorexia, nausea and vomiting, giddiness, muscle cramps and circulatory failure. It can arise insidiously and is undoubtedly responsible for a lot of minor neurotic symptoms in new arrivals in hot countries, dry or humid. Water depletion accompanies it and enables the actual concentration of sodium in the body fluids to remain normal and thus sometimes defy detection of deficiency. It is important to estimate the degree of haemoconcentration therefore when measuring electrolytes in the blood in such environmental circumstances. Urine estimation of sodium chloride, while decried in some textbooks of medicine, is in fact a useful and indeed necessary test for, in severe cases, salt excretion may actually have ceased. The details of the test can be found elsewhere but we would emphasise the importance of clean test-tubes and droppers, the latter particularly before each stage in the test. One word about the danger of acute heat exhaustion and collapse in infants and children with muco-viscidosis, happily very rare in Kuwait. These children lose large quantities of salt in their sweat and this diagnosis must be suspected when collapse without other reason occurs in hot weather.

Anhydrotic Heat Exhaustion: This is due to the failure of the sweat mechanism which leads to dry skin. In the rare case this may be congenital but these people do not survive in hot countries. It results essentially in the failure in the production or delivery of sweat and occurs when non-indigenous people have been exposed for some months to an inimical hot climate. Subjective prickly heat precedes the condition but the presence of many small vesicles deep in the skin may accompany it (Miliaria profunda). The systemic symptoms are similar to those of salt depletion but there may be burning and palpitations related to the hyperdynamic circulation and the breathing may be rapid, probably compensatory to increase heat loss via the lungs. When an individual is found with this type of heat exhaustion he or she should be removed to cool conditions as soon as possible because the temperature regulating mechanism has failed and the serious complication of heat-stroke is imminent. Permanent transfer to a cooler climate is mandatory.

Heat-stroke: That the Persian Gulf has long enjoyed a reputation for being as unsalubrious a part of the world as anyone could wish for is amply born out by the many descriptions of both its climate and its coasts: "Semi-burnt and sandy regions lie on all sides; not even a blade of grass relieves the aching eye-balls from the intense glare of the sand; the hot season, which continues for five months, is intolerable; existence then is almost unsupportable; the sun is so powerful during the day that it is almost certainly fatal to expose oneself in the least to its influence. I have seen men die in the utmost agony and raving mad, from exposure to the sun, after a few hours illness. When attacked with this brain fever few get over it, and if they do, their intellects are forever impaired" [5]. Elsewhere we have described the climatic conditions which are responsible for this dismal description and many before and since have written in similar vein. We will have occasion later to refer to the long term effects of acute heat-stroke but at this point it should be remarked that while much traditional mumbo-jumbo has accompanied the earlier accounts, it is only within the last thirty years that a clear picture of its cause, prevention and treatment has been generally acknowledged. Since the first somewhat primitive oil tankers began to ply the Persian Gulf, just prior to the 1914 war, with the products of the Abadan (Iran) refinery, those responsible for manning the ships have realised the high toll which heat-stroke has taken, yet it was all there before in British Naval history as our quotation above illustrated.

Even Nelson, who spent eighteen months in and around the Persian Gulf in 1775 and 1776, very nearly perished, for we have Southey's account [6] of his experience: "At this time his countenance was florid and his appearance rather stout and athletic; but, when he had been about eighteen months sailing out of Bombay and in the Persian Gulf he felt the effects of that climate, so perilous to European constitutions. The disease baffled all power of medicine; he was reduced almost to a skeleton; the use of his limbs was for some time entirely lost; and the only hope that remained was from a voyage home."

The factors which contribute to the escalation of the failure of the body to resist elevation of the body temperature do not arise from any one source and even yet they have not fully explained this very serious condition. In 1947 in the Persian Gulf, with the enormous increase in the numbers of tanker crews and a concomitant *heat-stroke* morbidity and mortality, there was an urgent need to apply such principles as were known and reduce what was fast becoming a serious load upon the limited medical facilities in Kuwait.

It was Guthrie [7, 8] who began to organise the facilities for acute treatment and practise the effective educational measures among the crews and tanker companies which over the next few years reduced the incidence of the condition among this special population to a trickle. Coincidental with the satisfactory waning of heat-stroke in tanker crews, arose a worrying incidence among the new immigrant population on land, particularly in the rapidly expanding urban areas of Kuwait, where living conditions, the absence of air conditioning and heat insulation, led to sometimes unbearable conditions at the height of summer. The Government hospitals were receiving over one hundred cases a year at this time and it was not until 1965 that the figure was brought down below this number [9].

Table 50. *Showing the number and percentage of patients in three categories of heat disorders in 1960 and in 1965*

Heat Disorders (body temperature)	No. (and percentage) of patients in 1960	No. (and percentage) of patients in 1965
Heat hyperpyrexia (over 105° F)	47 (35%)	2 (3%)
Heat pyrexia (100—105° F)	32 (23.7%)	16 (24.4%)
Heat Exhaustion (under 100° F)	56 (41.3%)	48 (72.7%)
Total	135 (100%)	66 (100%)

Heat-stroke follows the inability of the individual to balance the elevation of body temperature by mechanisms which effectively lower it. The range within which this can be successfully done is usually wide and undoubtedly is closely related to the acclimatisation achieved by the individual. Any deficiency of this, either by insufficient time in the climate, or by a loss such as follows illness, drugs and alcohol, will result, if conditions are sufficiently adverse, in the failure of heat dissipation, reduction, and finally cessation of sweating, and rapid elevation of body temperature to 43° C at which point death is inevitable. The unstable electrolyte and water metabolism preceding heat-stroke has led to awareness of both sodium and potassium abnormalities, the controlling effect and eventual breakdown of adrenal cortical hormones, in particular aldosterone and the termination in kidney failure [10]. Such aspects are beyond the scope of the present work but are of great importance in the continual battle to understand and prevent heat-stroke among a people who are entering a western industrial culture with its almost traditional contempt for the fallibility of the individual.

Two factors have an overriding importance in the prognosis of heat-stroke. First the height of the body

temperature at its zenith and second the rapidity with which this can be lowered. The first is dependent upon early recognition of the premonitory signs and the second upon effective though not excessive cooling.

The premonitory signs may be minimal with the rapid onset of coma and convulsions, but they are often preceded by weakness, headache, confusion and disordered behaviour, general diminution and then cessation of sweating, often polyuria and sometimes diarrhoea. The physical examination, beyond a dry hot skin, congestion or cyanosis, is remarkably normal except for the temperature. It is imperative to consider the possibility of accompanying disease precipitating the condition, particularly in those who may have recently been in disease endemic areas; the clinical picture of cerebral falciparum malaria may mimic or accompany *heat-stroke*. If the high temperature cannot be lowered within a few hours, the prognosis is grave. To this end in Kuwait great efforts were made when dealing with tanker crews, whose duration of high-fever was often prolonged before their ship was able to dock. Guthrie and McCracken [11] designed a rapid cooling table supplying brackish water at 4.5° C which was most effective in reducing rectal temperature within a short time, usually within 30 minutes. This table was supplied to all ship-berthing areas at the oil port of Ahmadi and continues to show its value although now its use is rare. Standard details of treatment can be found elsewhere, but we would like to emphasise the very great importance of early recognition for on this depends the success of treatment and avoidance of complications, for practical purposes neurological.

These complications in Kuwait have been reviewed by Salem [9]. He took two samples five years apart, which also illustrated the reduction in incidence during that period, from 47 patients in 1960 to 2 in 1965. His findings showed that neurological complications only occurred in the first group and his tables are illustrative of the clinical epidemiology. Complacency must not be allowed to creep in for exceptional conditions could arise and cause a tragic upsurge, for instance, civil disturbance and stoppage of electricity supplies; this was a real threat in June 1967 in the early days of the Arab-Israeli war.

To summarise the subject of heat effects, we emphasise two aspects of the problem; factors affecting heat control and action to be taken to avoid death.

Factors affecting heat control

Clothing: It is known that "The intimate association of clothing with the skin can profoundly affect the heat exchanges, and therefore the heat stress of any given

Table 51. *Giving details of ten patients who died or were disabled due to heat-stroke in 1960*

Initials	Age, in years	Sex	Nationality	Occupation	Temperature on admission	Duration of stay in hospital	Outcome
M.H.	40	M	Egyptian	Engineer	109° F	8 hours	Died
K.A.	22	M	Jordanian	Unemployed	105° F	24 hours	Died
A.R.	32	M	Irani	Baker	110° F	10 minutes	Died
S.G.	70	M	Pakistani	Tailor	108° F	15 minutes	Died
A.S.H.	25	M	Omani	Unemployed	108° F	12 hours	Died
K.S.A.	36	M	Lebanese	Mechanic	110° F	18 hours	Died
N.M.S.	19	M	Lebanese	Waiter	110° F	9 weeks	Quadriplegia and cerebellar ataxia
S.Kh.K.	54	M	Egyptian	Carpenter	109° F	18 days	Cerebellar ataxia
F.R.M.	43	M	Jordanian	Unskilled labourer	110° F	3 weeks	Mental deterioration
Sh.A.E.	28	M	Irani	Unemployed	109° F	4 weeks	Right hemiparesis

situation" [12]. The apparent paradox is that in hot dry deserts, those who are fully clad are at a physiological advantage over people wearing light short clothing: This is due to the effects of radiant heat. Moderately thick, full, white clothing which reduces gain from radiant heat and at the same time allows sufficient air movement against the skin to sustain evaporation, is the best. The traditional clothing of the Badu is unsurpassed in its ability to produce a good micro-climate but is not suited to work: If this has to be done, properly designed working clothing which reaches a satisfactory compromise is necessary.

Age: An increased likelihood of degenerative disease, particularly cardio-vascular, is the factor most affecting heat control here. Fluid and electrolyte exchange between extra and intra-cellular compartments and the rate of breakdown of cells with loss of potassium relate closely to the importance of further studies of heat control in the older age groups, for longevity is increasing in Kuwait.

Acclimatisation: A subject in itself but one probably not of significance in people who live in the area constantly. Even air conditioning does not reduce acclimatisation for moderate heat exposure once it has been gained, but recent illness can, for it will disturb the delicate hormonal balances which have been achieved.

Physique: Obesity or excessive muscular development which increases the ratio of body weight to surface area is a disadvantage. As has been shown in the case of desert animals, surface area is the important factor.

Exercise: This undoubtedly predisposes to the renal failure from muscle damage and myoglobinuria [10].

Drugs: Many people today are self-prescribers, from anti-histamines to aspirin: In addition doctors in the Middle East prescribe very freely. The effect of drugs may be to alter the thermo-regulatory control or cloud the mind under stress. The almost universal chewing of the leaf of the Khat plant, thought to contain stimulants, has not been investigated in relation to heat stress, for the habit is not widespread in Kuwait, but popular in other areas of Arabia.

Action required to avoid death from heat effects

In "Intolerable heat" conditions, life will be sustained only up to a maximum of three hours. Therefore time is of the essence and this means communication so that removal or rescue can be started at once. Vehicle drivers should adhere to a strict discipline when driving in the desert (Fig. 57).

XIV. Occupational Health

1. Epidemiology

The Department of Occupational Health of the Ministry of Health of the Government of Kuwait was established in 1965. Its first Director came to Kuwait after long experience of occupational health in the United Arab Republic. Prior to this, the Kuwait Oil Company had established criteria for recognition and control of occupational health problems within its own preventive medicine division. Coincidental with the Government's action the Company appointed its first full-time occupational health physician and technical assistant and over the next few years there developed close cooperation between Government and Oil Company, particularly within the field of the environmental health of petroleum workers. This was fortuitous but at the same time providential, for within a few years the Government itself entered the field of oil production with the establishment of the Kuwait National Petroleum Company, refining at Shuaiba and undertaking exploratory operations through its Spanish associates. Kuwait has presented a foretaste of industrial manning problems which are now being felt by other developing countries with small indigenous populations and large natural resources requiring expatriate labour forces to work them. With the control of immigration, a responsibility is placed upon the industries concerned to conserve the health and welfare of their labourers, to educate them, to improve their flexibility and increase their productive capacity. If these are not done, wastage becomes excessive, increased calls are made on Government for non-wages support or the necessity for repatriation arises with the embarrassment this brings with it when dealing with the less endowed countries from whence the immigrants came.

Controlled immigration into Kuwait has been officially encouraged, but uncontrolled illegal immigration has also flourished. It is from among these illegal immigrants, mostly Iranis, that a large proportion of the lower paid workers are obtained and for whom health and welfare services are denied, for they do not officially exist. While Government is constantly active to prevent the abuse of this source of labour, those who do arrive are exploited, a pattern now seen in the United Kingdom among illegal Pakistani and Indian immigrants. This abuse leads to a retention of the lowest standards of living for a sizeable minority.

For the known and registered indigenous and expatriate Arab, Indian, Pakistani and other workers, broad principles and aims were enunciated and have been steadily pursued. In the words of the first Director [1], "Occupational Health should aim at the promotion and maintenance of the highest degree of physical, mental and social well-being of the workers in all occupations: The prevention among them of the departures from health caused by their working conditions; the protection of workers in their employment from risks resulting from circumstances adverse to health; the placing and maintenance of the worker in an occupational environment adapted to his physiological and psychological equipment, to adapt both the work to the man and the man to his job."

It is inherent in the services provided that they should include all health problems whether occupational or general. This is a modern and practical approach, for with it must come a greater awareness among the medical and nursing staff of a nationwide state medical service of the influences and effects of the occupational environment on the general health of their patients. These then are the aims of the Department in Kuwait; ambitious but not unduly so and seeking to make the maximum use of available resources. The contact of the Department with other fields of preventive and clinical medicine has been very close.

The accomplishments during the first five years of the Department's existence have been considerable and are briefly summarised below [1]:

1. Issued regulations for the protection of the workers from occupational diseases and accidents to be followed in the places of work.

2. Standards of the sanitary facilities and general hygiene such as ventilation, lighting, lavatories, baths, washing facilities, rest-rooms, and places for keeping working clothes in special lockers et cetera, which all aim at the raising of the standard of health.

3. Issuing regulations for the protection of special groups of workers such as young persons, and women and issuing their certifications of fitness.

4. Regulations for the medical care for the workers and their supervision.

5. Standards of the pre-employment and periodical medical examinations.

6. Regulations for the limitation of hours of work for certain groups of workers, in dangerous trades; such as foundries, and jobs that expose the workers to dangerous gases or fumes.

7. Setting the standards for the maximum allowable concentrations of various air pollutants in places of work (MAC values); and the assessment of the concentration periodically.

8. Housing and nutrition of workers, canteens and restaurants.

9. Standards of the basic first aid equipment, plant clinic and the standards and qualifications of first aid officers, and supervising their work.

10. Training of the first aid officers and nurses.

11. Approving the designs of the new industrial establishments before construction to establish the hygienic methods of prevention of the occupational hazards before the work starts.

12. Issuing a schedule of the compensatable occupational diseases.

13. Issuing the regulation and procedures to be carried out in cases of accidents viz. notification and treatment, medical reports, rates or scale of permanent partial incapacities, and rules of medical arbitration in medical disputes.

14. Collecting data and statistics of accidents and of occupational diseases.

15. Carrying out routine periodical inspections to guarantee the fullfilment of the above mentioned regulations.

16. Training of medical and paramedical personnel working in the field.

17. Propaganda and health education in the field for the employers and employees, and the general population by all methods.

18. Planning the methods of rehabilitation of the workers and vocational training.

19. Establishing a library of books and magazines for references for all the persons working in the field.

20. Establishing contacts with the universities, institutes and other governmental or non-governmental organisations, and attending conferences both at local and international levels.

21. An occupational hygiene laboratory was established to carry out the various field surveys and research.

22. Dealing with problems of ambient air pollution from all sources whether industrial, natural or domestic.

Considerable assistance in implementing and reshaping the recommendations and requirements of the International Labour Organisation was obtained from the experience of other governments, in particular the United Arab Republic, the United Kingdom, and the United States of America. It is from the first of these that much of the statutory recommendations have been adapted. The schedule of compensatable diseases and the comprehensive list of the various permanent incapacities has now been in operation for several years. The financial compensation for these accidents and diseases is related to the daily wages of the worker instead of the old fixed sum known as the "dia"; in no case has it been less than the "dia" but increases as the wages increase. The estimation of disability has been based on the total body value instead of the old Kuwait "dia" system of judging this according to the value of the injured part of the body, which was often misleading to the courts and difficult to calculate.

Despite the nation-wide distribution of general medical services, the pressure upon these facilities prevented satisfactory medical supervision of the workers in industry, because doctors were unable to visit factories and carry out preventive and environmental work. It was soon realised that there was a great waste of time when workers reported sick to the general medical clinics and saw doctors who had little if any knowledge of their working conditions. The result was that they were indiscriminate in the advice they gave the patient. Large concerns were asked to appoint their own doctors and for smaller firms a special clinic was re-established in the centre of the industrial area in Shuwaikh. This has a historic relevance because over twenty years previously a clinic had been set up in the then "Labour Camps" area to avoid overwhelming the town clinics, a timely reminder to those authorities who still refuse to admit that people at work are still people, and may need to seek advice during working hours. Preventive work has lagged behind curative medicine due to the nature of the training of many of the private doctors appointed by larger companies and it has been the duty of the Occupational Health Division to urge a more practical and sensible approach while at the same time providing the expert guidance, advice and technical assistance in the form of industrial hygiene services. At the same time constant propaganda, health education and training have been pursued. The Director feels that a good start has been made bearing in mind that such services must take years to develop in the mixed economy of public and privately owned industry.

2. Petroleum and its Hazards to Health in Kuwait

In the book of Genesis [1] God instructed Noah to "Make thee an ark of gopher-wood; rooms shall thou make in the ark, and shall pitch it within and without with pitch." The Old Testament contains several direct and indirect references to the raw materials of petroleum which were translated variously as pitch, lime-pits and asphalt. These materials originated from the oil-permeated outcrops of rock, lakes of tar or bitumen and the more dramatic seepages of gas or liquid oil; perhaps the best known of these were in northern Mesopotamia, at modern Kirkuk, familiar to many of us as "The burning fiery furnace" [2]. Another great religion which had fire as the symbol of the pure and the good was associated with these seepages, this time at ancient Shushan or Susa in Persia, now Shoosh in the Iranian province of Khuzistan. The ruins of the temples of the Zoroasteans can still be found located where the many smaller seepages were known to have occurred and which eventually led to the discovery of the oil riches of Iran. Pliny [3] vividly describes the fires at Susa: "... likewise at Susa at the white tower, from fifteen apertures, the greatest of which burns in the day-time". These same petroleum springs had been noted by Herodotus [4] centuries before when he described the natives of the place baling out the oil from goat-skin buckets at the end of a long pole, a method of oil and water-recovery which has come down unchanged through the centuries.

Bitumen ponds or lakes had been known in the desert of Kuwait from earliest times: dare we imagine that it was pitch from the Burgan field that was used for the sanitary engineering dating from the third millenium before Christ which Sir Leonard Woolley [5] found during his excavations at Ur of the Chaldees? Ur, the modern Telnoughaiyur, on the Baghdad to Basra railway, lies not more than 160 kilometres by desert camel from these bitumen lakes. Neolithic arrow heads and scrapers have been found in their vicinity suggesting very early exploitation. Yet in more modern times this must have lapsed, for we have no evidence in Kuwait of the village industry of oil production carried out elsewhere in the valley of the Tigris described by Longrigg [6]: "... sometimes by collection or skimming potful by potful, from a surface trickle of the dark fluid. Rough galleries were made to improve access and shallow holes were dug by hand. Bitumen collected from the ground was taken on donkey or camel-back to a place of melting."

Fractionation of crude petroleum, as we know it today, was certainly being carried out on the western shore of the Caspian sea, when Geoffrey Duckert [7], agent of the English Muscovy Company, visited Baku in the sixteenth century. The only commercial product was what we know as kerosene, used principally as an illuminating oil. It was the more general introduction of this oil which gradually caused the decline of the great whaling industry upon which supplies of domestic oil had, up till then, depended.

The modern techniques of oil refining had to await both the development of organic chemistry and engineering and the design of apparatus far beyond the crude laboratory scale; they eventually became commercially fruitful almost simultaneously in Russia and the United States of America by the middle of the nineteenth century. It was, however, the development of the internal combustion engine towards the end of the century which sent prospectors expectantly into those areas long known to have seepages. The problems were not easy of solution, as instanced by the story in Iran [8, 9].

Search for sources of crude oil began in Kuwait in 1934 but here again the vicissitudes of desert, political and commercial intrigues and ultimately the 1939—45 war, together postponed the commercial production until 1946; even then this was limited by the shadow of the great production capacity of the oil-fields and refinery

just to the north in Iran. But "It is an ill wind that blows nobody good"; the political outburst, which led to the expulsion of the Anglo-Iranian Oil Company from Iran in 1951 followed by the breakdown of oil supplies and refinery capacity, served Kuwait well, for sudden demands were placed upon its production which were able to be met and have since been followed by continuous expansion.

Table 52. *Use and destination of Kuwait petroleum products*

Product	Use
Propane	Export to Japan
Butane	Export to Japan & local market
L.P.G.	Export to Japan
Premium gasolene	Local market
Kuwait light distillate (90% Hexanes)	Local market
Gas oil	Local market & export
Aviation kerosene	Local market & export
Kerosene	Local market & export
Marine deisel oil	Export & bunker
Fuel oils	Export & bunker
Bitumen	Local asphalt plants

Adapted from:
The Economic Development of Kuwait. Baltimore 1965, p. 198. Library of Congress Catalogue No. 65–111664.

Burgan oil has a great disadvantage which it shares in differing degree with crude oils from other sources in the Middle East and elsewhere, for it has a measurable and indeed embarrassing concentration of sulphur as hydrogen sulphide, varying between 0.1 and 0.4 per cent. With the legislation in many industrial countries controlling the quality of effluent permitted, the emission of the products of combustion is carefully defined with particular emphasis on sulphur. This has restricted the use of Kuwait crude oil in some European refineries; elsewhere those countries which greatly rely upon it, in particular Japan, have had to plan expensive desulphurisation plants (see Table 52).

Petroleum is the most versatile of all forms of potential energy; its fractions are utilised as fuel, propulsive force, lubricants, illumination and for their medicinal properties. More recently we have seen the establishment of the massive petrochemical industries which feed upon the sophisticated products of petroleum fractionation and now its potential as a source of food energy and growth in domestic animals is being developed. But relatively few of these uses are effected in Kuwait where daily, millions of cubic feet of natural gas are still burnt to no other purpose than that they are excess to requirement: Yet other quantities are injected back into the porous rock bed to maintain reservours and sufficient pressure, to achieve a flow of oil without the necessity to pump it up as is done elsewhere.

Our purpose is to examine the real and potential damaging effects of petroleum production and its derivatives upon the people in Kuwait, for they live in close and continuous association with them. They see the burning flares in the evening sky, clouds of black products of combustion hang over their heads, they smell the sickly stench of mercaptans, the automobile is their symbol of status, and if they give any thought to it, they could not exist as they do today in comfort without the dozens of oil tankers continually anchored off the coast, awaiting their turn at the loading jetties but contaminating the waters of Kuwait's shore with waste oil. Perhaps the greatest danger from petroleum today in Kuwait, is that, as sure as night follows day, if during the hot season, oil and gas production were removed by some action of God or man, a very large proportion of the people of Kuwait would die if help was not brought within hours. Petroleum helps to provide their water and their comfort and indeed for most of them their very reason for being in Kuwait at all, although few are directly involved in the industry. Not many people would survive the summer temperatures after they have become accustomed to the artificial environment they have created, if petroleum gas was suddenly cut off; a threat not entirely fanciful as instanced by the strike of oil workers during the six days Arab-Israeli war in July 1967. Day temperatures were up to 49° C (120° F) in the shade, and production was maintained by a handful of responsible Arabs and expatriates.

Not all the direct effects of the *use of petroleum products* are unfriendly *to the human body*. For centuries they have been in use as medicines, vermifuges (drugs for purging the bowel of worms), vapours inhaled for consumption, as inunction for painful muscles and joints and to rid man and domestic animals of ticks and lice. Perhaps the most famous of commercial crude oil preparations sold under a label was Seneca Oil, which was introduced in the United States of America in 1859. It was claimed as a universal panacea and survived as a household remedy for several decades. The beneficial nature of the crude material remains unproved and it is on the refined products that pharmacy relies, in specific fields such as ointment bases, vermifuges and parasiticides; yet even these have largely been superceded by synthetic compounds. The Badu today is more likely to purchase a modern remedy in the suq for his animals, or obtain the advice of the Government Veterinary Department than resort to his age-old traditional recipes, for with his pickup truck which is fast replacing the camel, a ride to the city is but little chore.

Elsewhere we have identified the number of people actively engaged in the petroleum production and refining industries as a very small proportion of the total work force in the country. Many of the hazards which were almost a daily occurrence in the early days in America and Iran have been eliminated. Nevertheless, the total population of Kuwait is daily exposed to the effluents of petroleum production and their long-term effects may not be apparent for many years.

In the manufacturing of petroleum the dangers most to be feared are those not foreseen: In an industry constantly developing, potential dangers are very real as exemplified by the introduction of the hydrogenation process in the new Government refinery at Shuaiba, in which liquid hydrogen in large amounts is continuously circulating (Fig. 58). In the history of the industry there have been many tragic experiences, several of these in recent times in the Middle East. In the space of fifteen months in 1964—65, there were five major oil tanker explosions at the head of the Persian Gulf involving Kuwait, Iran and Saudi Arabia. While the accompanying Table 53 looks impressive it by no means represents all the potential hazards of the petroleum industry but rather the day-to-day exposure met with by a relative handful of workers in Kuwait (Figs. 59 and 60).

Table 53. *Hazards of petroleum industry*

The hazards	To whom	Site where used
A. Acids		
Aluminium	E[a]	Catalyst
Ammonia	E[a]: Public	Refrigeration plant
Arc-eye	E[a]	Welding
Aromatic amines (mono-ethanolamine)	E[a]: Public	Nitrogen plant
Asbestos	E[a]	Insulation of pipes and roofing sheets
B. Barytes	E[a]	Special muds for drilling
Bentonite	E[a]	Special muds for drilling
Benzene	E[a]	Petroleum laboratory; maintenance of Platformers (Refinery)
Beryllium compounds	E[a]	Special non-sparking tools
Bitumen	E[a]	Splashes from Bitumen plant
C. Cadmium	E[a]	Catalyst
Carbon disulphide	E[a]	
Carbon tetrachloride	E[a]	Electrical workshops
Caustics (NaOH)	E[a]	
Chlorine	E[a]	Water purification
Chlorinated hydrocarbons	E[a]: Public	Insecticides
Chromates	E[a]	
Cobalt	E[a]	Catalyst and cleaning of refinery plant
Cyanide	E[a]	Laboratory
D. Diving	E[a]	Pier and sea-line maintenance
E. Explosive accidents	E[a]: Public	Tankers, pumping stations, pipelines, laboratories
F. Fluorides	E[a]	Fluoridation plant, welding
Fluorinated hydrocarbonates	E[a]	Refrigeration gases
H. Hydrogen sulphide	E[a]: Public	Gathering centres, Refinery, (Hydroban & Platformer), Gas injection
I. Ionising radiation	E[a]	Plant survey, metal fabrication, pipelines, drilling operations
1) Ameridium	E[a]	
2) Cobalt	E[a]	
3) Iridium 192	E[a]	
4) Radium-beryllium	E[a]	
L. Lead, Litharge, Wet plumbite, Tetra-ethyl lead	E[a]: Public	Refinery
M. Mercaptans	E[a]: Public	Refinery
Mercury	E[a]	Refinery laboratory
Methanol	E[a]	Refrigerated pipelines (natural gas)
Monoethanolamine	E[a]: Public	Nitrogen plant
Molybdenum	E[a]: Public	Catalysts in petroleum fractionation and cleaning the plant
Metal fume, phosphor-bronze, brass, zinc	E[a]	Main workshops, metal spray-guns
N. Nickel	E[a]	Catalyst
Nitrogen	E[a]	Nitrogen plant, liquid nitrogen in laboratory
Nitro-aromatics	E[a]	
Nitro-paraffins	E[a]	
Noise	E[a]	Gas separators, gas injection, turbines
Nitric acid	E[a]	Petroleum laboratory
O. Ozone	E[a]	Welding
P. Phosphorus (organic)	E[a]	Insecticides
Platinum	E[a]	Catalytic process, cleaning of plant

[a] E = Employees

Table 53 (continued)

The hazards	To whom	Site where used
S. Silica	E[a]	Sand-blasting, Batching and crushing, furnace and boiler repairs
Sulphur dioxide	E[a]: Public	Refinery
T. Toluene	E[a]	Engine test laboratory
Trichlorethane	E[a]	Cleaners
Trichlorethylene	E[a]	Solvent
V. Vanadium	E[a]	Boiler and tube scaling

There has been little trouble experienced with any of these substances or processes because of the careful monitoring carried out in the field and the periodic examinations of personnel exposed.

However, with the expanding total economy and the efforts of the oil companies to rationalise their own operations, there has been a steady trend to reduce manpower and to rely upon contract services. This has led to problems of health education among the contract labour who are not exposed other than occasionally and often do not have the ability to understand the significance of safety measures. A typical example was when the tetraethyl lead tanks in the refinery were required to be cleaned by removing the sludge: A relatively simple operation requiring air-hoses and breathing helmets and a strict discipline to avoid any worker being overcome with the toxic vapours. Yet in 1962 several men were overcome, fortunately not seriously, because of the failure of the contractor's supervisor and the company official to ensure that the code of practice was adhered to.

Three examples will now be given to show the variety of hazards which can be experienced, although rarely.

At a new gas-injection plant there was a known hazard of hydrogen sulphide. Medical and safety advice had been given that this should be dealt with by blowing off from a high stack, for it would disperse with the wind and not reach the ground to lie in pockets, where it could be dangerous by its acute poisoning effects and explosive capability. The advice was not taken, the excess hydrogen sulphide was led off by a pipe-line at ground level into the desert outside the injection plant area. This would have been of little consequence had not one day a Badu truck, loaded with family and house-hold goods, driven past the plant and through a pocket of gas. There was a violent explosion, the truck was wrecked, three of the family were killed and some very embarrassing questions had to be answered. The electrical system of the truck's engine had ignited the gas.

An example of a similar type of occurrence was recently reported [10] from the United States and serves to illustrate the hazards of our modern technically orientated life "to people not directly engaged in the manufacturing or scientific processes".

"In a bazarre accident at Cape Kennedy, Florida, three cars driven by security officers burst into flame yesterday as they passed throuh an area only 1,100 ft. from the Appollo 13 moon rocket.

Oxygen vapour from storage tanks near the launch pad had apparently collected in a pocket, and when the cars entered it their engines caught fire. The drivers ran to safety.

The National Aeronautics and Space Administration said that the accident was unlikely to affect the launch of Appollo 13 on April 11. If the fire had ignited the liquid oxygen in the storage tanks, the rocket might have been destroyed."

A tanker was loading crude oil at the jetty when there was a series of explosions beginning in the officers' quarters under the bridge, amidships. Some twenty men were either killed at once or died following severe flashburns. Fortunately the breeze was off-shore. The ship was towed away from the jetty at no little risk to the skeleton crew provided by the Oil Company, while firefighting continued. A naked electric light bulb was the cause of the initial explosion. The responsibility at that time for safety precautions on ships at sea or at the jetty was with the captain, and naked electrical points should not have been exposed to potentially explosive gases arising from the crude oil being loaded. In addition the air vents of the loading tanks in the ship were defective. This episode exemplified the need to:

1. Ensure safety precautions in every ship lying at the jetty.

2. Crew should wear anti-flash clothing to prevent whole body burns due to the minimal clothing worn in the hot humid climate.

3. Have a major distaster plan available to cope with episodes of this nature. Fortunately at this time such a plan was in operation.

The final episode was the loss of an Ameridium "bomb", a powerful sealed source of radio-active energy with a half-life of 900 years, some 12,000 feet down an oilwell. The problem was twofold: First to recover the "bomb" by grappling it with a special tool and second, in the event of fracture of the metal casing to avoid the possibility of radio-active contamination of the oil workers attempting to do this. A health physicist and radiation protection team were flown out from Britain, and continuous monitoring of all drilling mud pumped from the well together with rigid control of the environment at the well-head were instituted. With good luck and good technique the radio-active source was recovered intact and there was no leakage of radiation either in the well or at the surface, but the tale might have been otherwise.

The additive effects of sun radiation and the carcinogens from crude oils on the skin have not so far led to an increased risk of skin cancer in Kuwait, and studies carried out between 1958 and 1963 in the adjacent Saudi Arabian oil fields have not indicated a higher incidence among expatriate Americans than is normally found at similar latitudes in the United States [11] (Fig. 61). The potential carcinogenic effects of Kuwait crude oil have recently been studied [12]. Actual direct contact with crude or its products is very limited; this together with the habit of the Arab oil worker, many of Badu stock, to cover themselves against the sun greatly reduces the irritative effect of the latter, which could potentiate the former.

No evidence of increased cancer prevalence in other organs, particularly the genito-urinary system, exists, but it is early for such effects to show. A cancer registry has not yet been set up by which patients who may have been exposed to carcinogens at some time in their working life may be identified.

Crude mineral oil contains many different hydrocarbons and related substances. These can now be physically separated and identified; recent work [12] has, for instance, isolated forty-five pure compounds not previously recognised as constituents of uncracked mineral oils. Most of them have turned out to be derivatives of polycyclic aromatic hydrocarbons and aromatic heterocyclic systems; some have given clear evidence of the presence of anthracene derivatives, such as 6,7-dimethyl-1,2-benzanthracene, a well known though weak carcinogen for mouse skin, but more potent in rabbits. Recently 4-methyl-1,2-benzanthracene and 7-methyl-1,2-benzanthracene have been isolated and identified. There is now a strong suggestion that "some of the fractions of Kuwait oil which are themselves only weakly carcinogenic may enhance the activity of the standard carcinogen" [12]. These may be present in the uncracked mineral oil. No single carcinogen has been discovered so far, but "groups of compounds related to the known polycyclic carcinogens are present in the oil, and it is likely that the carcinogenic activity of the oil is due to the combined action of groups of closely related components that are difficult to separate" [12]. Again "the presence of large amounts of sulphur compounds may also be important, for a number of polycyclic sulphur compounds have been shown to be powerful carcinogens" [12]. As yet however, there is no evidence that pentacyclic benzopyrenes, in particular 3,4-benzopyrene, which is strongly carcinogenic, are present in Kuwait crude oil, although this may not be true for higher boiling fractions after refining.

This work should stimulate an awareness in Kuwait of the *long-term effects* of crude oil and petroleum products *on the skin and internal organs*. There is a fruitful field, we would think, for investigations to be done now on human and animal tissue concentrations of these substances analogous to the tissue residue surveys carried out elsewhere for DDT and the chlorinated hydrocarbon pesticides, for the population can be very easily sampled. Interest, if not alarm, has recently been generated from two further directions in relation to petroleum products hazards; while not solely applicable to Kuwait they are worth considering in the total context of petroleum products and carcinogenesis.

Mineral oil lubricants have been under increasing suspicion as harmful contaminants of processed food. These lubricants are white mineral oil, odourless light petroleum hydrocarbons, petroleum wax, white petrolatum and distillates such as heptane, hexene, kerosene, mineral spirits and naphtha. They are used as lubricants of food processing machinery and water pumping plants, as protective anti-rust film; as release agents on gaskets or seals of tank closures, as defoamants in, for instance, beet sugar processing and in the extraction of spices and curing pickles. In some countries laws have been passed to control and limit the use of mineral oil products for these uses [13].

Related again to food contamination is the perhaps more important and perplexing problem of the (aliphatic) amines, the nitrosamines, which are known to be powerful carcinogens [14]. Particular interest has arisen recently because of the suspicion that they can be formed "in vivo" by a reaction between nitrites and secondary amines, the nitrite having been deliberately added to the food (meat) as a preservative and colouring agent. But natural human nitrate ingestion occurs from plant food; storage and preparation reduce this to nitrite and reaction may then take place in the stomach with secondary amines, known to occur in fish products cereals, tea,

tobacco smoke and some alcoholic liquors. It is very possible that relatively high concentrations of these nitrosamines could be found in the environment of Kuwait arising from the incomplete combustion of petroleum products. Small quantities of these products are found in the oil refinery, particularly in the nitrogen plant. The institution of a cancer registry would thus be a valuable epidemiological tool and should not be delayed. Reference has already been made in Chapter XII, 4 to the potential danger of amines acting as triggers for the development of acute haemolytic anaemia in Glucose-6-phosphate dehydrogenase deficiency.

Finally, again in the oil refining process, potentially dangerous dusts arise from the platinum, cobalt and molybdenum catalysts during cleaning operations of the hydrobon and platformer, catalyst regeneration and scouring of the metal tube bundles. No recognisable effects, such as the skin and respiratory symptoms experienced elsewhere in platinum refining plants have been identified, but a potential risk exists. It is now suggested that such trace elements, either in excess or deficiency may link closely with exposure to the nitrosamines in the human body and potentiate the risk of cancer. Perhaps a more serious risk exists to cattle who may be exposed to certain dusts, in particular molybdenum, when carried by wind into their pasture areas [15]. As yet no such problem arises in Kuwait, but with agricultural development the possibility should be foreseen.

3. Secondary Industries

The accompanying table of secondary industries (Table 54) is impressive but belies the actual contribution to the gross national product that it actually has. In addition, only a small proportion of the total labour force is engaged in this field, amounting to 9 per cent of those employed at the time of the 1965 census (see Tables 5 and 6). In Chapter IV we have already reviewed the economic aspects of the oil industry in rela-

Table 54. *Secondary industries*

Agriculture	Furniture
103,316 Kg milk	
1964 { 262,842 eggs	Gases — industrial, medical
92,993 poultry	
550 hectares cultivated	Goldsmiths
Aircraft maintenance	Ironwork:
	Industrial, Ornamental
Asbestos pipe and roofing	Lime production
Automobile repair	Pearling (almost extinct)
Brickmaking	Printing & Publishing
Catering	Road making
16 Hotels	
Industrial contracts	Salt, Chlorine, Caustic Soda
Construction,	Shipbuilding
including cement	
Dressing of hides and skins	Steel, strip pipe making
Dressing of stone and marble	Textiles
Electricity	Tent and Sail making
Engineering (light)	Water production
Fertilizer (petro-chemical)	
Fishing (trawling)	
Flour milling	
Food:	
Ice cream	
Soft drinks	
Powdered milk (imported)	
Fresh meat	

tion to the remarkable small contribution it makes to the labour force.

It has been difficult to estimate the intensity of occupational environmental health problems from the scanty information available, largely due to the scattered and irregular pattern of industry, but three classes can be identified and the Government occupational health services have been concentrating their attention upon these.

Table 55. *Hazards of secondary industries*

Dust	Infection	Chemical toxicity
Asbestos	Slaughter houses	Chlorine
Stone dressing	Dressing of hides and skins	Medical gases
Cement blocks and tile making	Catering Industry	Engineering (solvents)
Wood-work and furniture making		Agriculture (pesticides)
Kuwait flour mills		Printing
Lime production		
Brick-works		
Textiles		
Metal-work		

Little more than identification of the major hazards of *asbestos* fibre, stone-dressing and pesticides has been attempted to date, concentrating upon the improvement in environmental control, personal protection equipment and individual monitoring by clinical examination and radiography. Nevertheless, the establishment of the asbestos plant in 1965 was an object lesson in the need to assess hazards before an industry is established; initially there was virtually no control of asbestos dust within the plant itself or of its emissions. Within a year criticism of the operation was widespread but the damage to susceptible people exposed, both in the factory or environs has perhaps yet to reveal itself; already one of the characteristic pulmonary effects, intrinsic asbestosis, has been seen, but carcinoma of the lung, pleural plaques and mesothelioma have not yet been reported. The mistakes have not been repeated in a second asbestos plant recently established. Two traditional Kuwait industries, known elsewhere to be related to chronic late effects are those of goldsmiths (jewellers) and pearling, the former with lead and mercury poisoning and the latter with destructive lung disease or emphysema. However, no evidence of specific illness of goldsmiths or jewellers has presented in Kuwait, although that is not to say it does not occur. Pearling on the other hand has now almost ceased and only relatively few men can be identified as having been engaged in the industry over sufficient years to have caused lung damage. Nevertheless, between 1961 and 1968, one of us identified three men in the 40—50 age group who presented with erythraemia, which is an excess of red, oxygen-carrying cells in their blood (Table 56). None had any degree of emphysema sufficient to account for a secondary erythraemia, due to chronic anoxia from respiratory failure. As these may be unique observations fuller details are worth recording. No deductions as to the relation with pearling can be drawn, but it is known that these men, particularly when young, were subjected to both considerable physiological and psychological stress and it is possible that in a proportion, the former might have stimulate chronic red cell overproduction in the bone marrow from the demands made by the anoxia. The psychological stresses have been emphasised by Dickson [1] when describing the chronic load of debt incurred by these men and the debased social quality of their family lives, particularly the prostitution of their wives to supplement the family income. With the demise of pearling and the vastly improved conditions of life in Kuwait, such problems happily no longer exist.

Training for industry in Kuwait is being carried out in the Government Technical College and results of this have begun to show in that an increasing number of young people born in Kuwait or who have resided there since childhood are now available for industrial work. Still the number of urban Kuwaitis is limited, but an appreciable intake of Badu Kuwaitis is beginning to redress the balance. The wealth obtained from petroleum should provide the means to set the economy on a sound industrial basis: This has been appreciated by the authorities and their advisers, but development is restricted by manpower on the one hand and lack of sufficient markets on the other. The assistance that Kuwait is giving to less well-endowed States further down the Gulf in the fields of education, planning and public health, should eventually provide a local market for Kuwait-made goods but the competition will be fierce, requiring heavy Gov-

Table 56. *Three cases of industrial disease?*

Name	M.S.A.	M.H.M.	A.A.
Age	45	39	50
Nationality	Omani	Kuwaiti	Kuwaiti of Irani origin
No. of years pearling	7	6	15
Smoker	4 cigarettes daily	Hookah (Argila)	Hookah (Argila)
Presenting complaint	Pain in left loin	Routine examination	Routine examination
Emphysema	Nil	Nil	Nil
Enlarged spleen	Nil	Nil	Nil
Peak flow metre (breathing test)	460 litres/min. (normal)	—	—
Haemoglobin (normal 100%)	134%/19.8 Gm	137%/20 Gm	143%/20.8 Gm
Packed red cell volume (normal 45%)	59%	65%	61%
Recognised accompanying disease	Left renal calculus and microscopic haematuria.	Craniostenosis of skull. Hyperostosis frontalis. Normal pituitary fossa.	Nil

ernment subsidies to achieve competitive prices. The effort to do this will however reach towards the essential goal of diversifying Kuwait's economy and creating a wider degree of interdependence among the Arabs of Arabia.

The academic, technical and *health education* required to enable the people of Kuwait to meet the demands provides a golden opportunity for those responsible in these fields to come together and inject into the educational programme first an awareness of what the future should be and second the means by which it can be achieved. This should begin with the secondary schooling and teach-ing the children the basic factors about the conservation of resources and manpower, the relation of good health to production and the satisfaction that sound training brings to the breadwinner and his family. Perhaps more so than is done in the West, the family should be included in the general consideration of occupational health. Here the experience of the great companies who have developed the petroleum resources can be useful in planning such an integrated system. The worker in the setting of his family whose care is also the responsibility of the industry has long been the health policy of these organisations.

XV. Air Pollution

Despite the fact that Kuwait is still strictly speaking an arid zone country or indeed because of it, its atmosphere has long known *pollution from particles of sand and dust,* but owing to their relatively large size, greater than five microns in diameter, they have exerted no more than a nuisance effect. This explains the absence of silicosis in the animal and human dwellers of the deserts who breathe the silicotic particles throughout their life. The *dry sand or dust laden air* is, however, very irritating to the upper respiratory tract, causing dryness and crusting leading to universal habits of picking the nose, hawking and spitting. But the desert dweller still follows a sound custom which is now being discarded in the villages and towns, for during hot windy weather he uses the loose ends of his Khaffiya or headdress to wind round the lower portion of his face and neck as a protection: This has the two-fold effect of filtering coarse particles from the air he breathes and conserving the moisture from his lungs to increase the very low humidity of the air.

The larger particles of sand are trapped mainly in the upper air passages but those which do get down into the bronchi are soon expelled and do not became deposited in the lungs. In the past the irritation caused by sand has not led to any recognisable respiratory illness per se, unless there is some synergistic effect with the tubercle bacillus. But with the enormous increase in the habit of smoking unfiltered cigarettes over the past few years the additive effects have given rise to acute and chronic bronchitis as well as exposing the mucosa to the risk of respiratory allergy. It should be remembered that prior to the coming of oil in 1946 only a few better-to-do citizens could afford the western habit of cigarette smoking and the rest adhered to the traditional hookah which, although not aesthetically acceptable to the westerner, does filter the smoke through water, but with what advantage is not known.

The consumption of *cigarettes* today in Kuwait is enormous; imports for 1965—1966 were 6,000,000 K.D., a per capita figure of 15 K.D. It will remain to be seen whether there is a startling rise in the incidence of chronic bronchitis and carcinoma of the lung which has occurred in other countries. These diseases are already being diagnosed in the Kuwaitis but whether this is due to an overall age increase in the population or to a real increase it is not at present possible to say.

The *plan of development in Kuwait* reveals the coming into existence, already well on the way, of a major conurbation stretching along the coast for some thirty miles and with a depth of five miles. The build-up of industrial and automobile pollution along this narrow strip is inevitable though not necessarily an every day hazard. We will recall Professor Dudley Stamp's words, "Danger lies not in the average or mean conditions, but in the exceptional circumstances". The analogy of this remark with the holocausts which raged through the pages of the Old Testament springs at once to mind, particularly as they occurred in the same part of the world of which we are talking. Anyone who has witnessed any of these events will appreciate the terror which struck at the hearts of those simple folk. We recall a morning in 1965 when, upon waking, all that could be seen outside was a blackness tinged with red, with almost no visibility; this was due to an extraordinary storm which had lifted masses of grey and dirty sand and dust into the air and given a ceiling of only a hundred feet. Reflected from the layer of thick dust were the flames of the many burning oil flares in the desert: The whole scene, with the howling wind, was terrifying to anyone who did not immediately understand this unusual combination of circumstances. The rumour and panic that can run through a people may strike deep into their character, for such happenings in Kuwait do indeed occur every few years and until recently were the means by which historical time was recorded.

Examination of the plan of Kuwait residential and industrial development reveals some rather alarming possibilities when taking into consideration such natural variants as wind direction, humidity and atmospheric ionisation for already into the air of this narrow belt is poured daily the waste products of the petroleum and growing petro-chemical industry. Added to these are the exhaust dusts from the asbestos industry, the products of incomplete combustion of industrial, domestic and automobile gas and petroleum units, which produce sulphur dioxide, polycyclic aromatic hydrocarbons the best known example being 3, 4, benzpyrene, carbon monoxide, nitrogen oxides, aldehydes and unburnt hydrocarbons, fluorides, arsenical vapours, lead and a variety of other identifiable pollutants.

In Kuwait the heavy unburnt particles in the black smoke from the burning of crude oil during separation of gas and lighter oils are often seen to form a flat layer of black cloud quite close to the ground, due to the ceiling of warm air causing a "temperature inversion". It is this phenomenon elsewhere which, when combined with high humidity such as in fog, causes a "smog", a form of aerosol mist which can penetrate deep into the lower respiratory tract: The admixture of heavy quantities of sulphur dioxide introduces a very irritating effect upon the bronchial mucosa. Whereas the above is more likely to happen where the temperature and sunshine are not high, in industrial areas with high sunshine levels there may be photochemical changes which alter the nature of the pollutants, and the carbonyls and nitrogen oxides resulting from the cracking of petroleum compounds interact to form compounds of the peracetyl nitrite type and ozone: The "smog" then becomes both irritant and oxidising.

As yet neither of these "smog" phenomena has been observed in Kuwait, but all the participating agents are already there and given sufficient concentration and climatic conditions could descend upon an unsuspecting population and reek havoc among the very young and the old.

The map (Text Fig. 25 on the back of Map 3) identifies the potential factors which could lead to trouble.

There is great need for the authorities *to recognise and legislate against these potential dangers* for once the population has been lulled into complacency a great deal of education will be needed to modify their tacit acceptance. I would quote from a recent review [1] which epitomises our apprehension: "Our knowledge of the toxicity threshold of chemical pollutants found in the air of cities and industrial environments is still, in the majority of cases, in an embryonic stage. We are particularly vague as to the insidious long-term effects on the health of populations exposed to polluted air. This lack of certainty is all the more disquieting to toxicologists and public health authorities since certain pollutant substances have been found to be definitely harmful in even minute quantities; this has, at least in a number of cases, led the authorities in various countries to impose regulations aimed at reducing the emission of these substances e. g. the "Clean Air Act" in the United Kingdom and the French law on pollution of the atmosphere".

Kuwait has a problem in that the economic and social effects of legislation might become too heavy a burden, for the limitation of black smoke pollution, while already controlled, is bound up with the life-blood of the country and the emission of automobile exhaust is inextricably linked with almost the only outside leisure activity of the modern Kuwaiti, which is to drive fast on the network of 1,200 km of new roads.

It will be seen from this map (Text Fig. 25 back of Map 3) that industrial pollution can sweep over the residential areas of Kuwait from the north-west and petro-chemical effluent can do the same from the south-east; it is this latter which is probably the greater hazard because of the heavy humidity the wind brings with it, maintaining the pollution in suspension and solution. The asbestos dust which sweeps across the southern residential areas from its origin in the one very large factory could be a long-term danger: There is also the secondary eroding effect of the sand-laden north winds on asbestos

roofing, the use of which is widespread in Kuwait. Only the future will tell whether the dreaded mesothelioma tumour will occur. It would seem justifiable to monitor the air-borne solids by means of the Hirst precipitator, which had previously been in constant use for some years to measure and identify air-borne moulds and pollens in southern Kuwait. A report should soon be available concerning wind-borne asbestos fibres trapped from the air over Ahmadi.

In 1965 Kuwait took a good look at the problem of air pollution with the *help of the World Health Organization* and the conclusions were that "There did not seem to be a very great problem of allergy in that the manifestations have stimulated little serious investigation of cause, prevalence or definition apart from the work reported by Wilkinson, which is a fair statement of the problem in Ahmadi" [2]. Wilkinson [3] had shown "That in a country previously free from allergy, the importation of vegetation consequent upon a plentiful supply of water and associated with climatic conditions ideal for *pollen production* and dissemination is responsible". He recommended the exclusion of known potent hay-fever producing plants as far as possible from the vegetation when arid regions were being developed. He also considered the effect of air-conditioning and suggested the use of filters in air-cooling units.

Wilkinson's results, following the attempt to desensitise patients, were not clear although he felt that the pollen of *Prosopis spicigera* (mesquite, algarooba or honey locust) was of considerable significance; many people of all races were shown to have positive skin tests. In fact, following these suggestions there was considerable lobbying to have the *Prosopis tree* destroyed. This overemotional reaction was fortunately not productive, but the next winter (1963) a considerable number of *Prosopis* trees in Kuwait were fortuitously destroyed by frost. This also coincided with the departure of Wilkinson from Kuwait which removed the main stimulus to further studies, the importance of which was stressed the following year by Lawther [2]. Today it is hoped that the collection of vital statistics, so important in obtaining a balanced view, will soon provide the epidemiologist with his ammunition. Nevertheless it is the present writer's opinion that up until 1968 the overall morbidity effect was minimal although for a few people there was considerable discomfort (see Text Fig. 26 on the back of Map 3).

Davies's results from the Hirst precipitator [4], of which previous mention has been made, have clarified some of the rather confused thinking occasioned by personal and clinical impressions. He identifies the pollen of *Prosopis spicigera* as an important cause of pollinosis despite the fact that they were seldom found in the spore trap positioned as it was above a building and open to the winds of the desert. In fact, the large, round, heavy spores of *Prosopis* are shed in clumps underneath and close to the trees which were planted and are used for shade. Thus the spores only occur in high concentration near the trees. It is interesting that more than one reliable witness has identified the acute onset of his or her symptoms when standing or sitting under *Prosopis* trees. Recommendations therefore to individuals to avoid the trees in the April/May and November/December flowering periods should be made, and susceptible people should be housed as far as possible upwind of the *Prosopis* trees, that is north-west of the planted areas.

As far as air-borne spores are concerned, the pollen of *Chenopodiaceae* was predominant to the extent of 66% of the total. They derive from the salt-bush *(Salsola baryosma)* and thorny scrub *(Cornulaca leucacantha)* which were originally sown to bind the sand, and which also occurred naturally near the sea-shore. The original object of their introduction into built-up areas has been very successful but the secondary effect is now suggested by Davies to be a cause of allergy in Kuwait. He quotes Frankland in saying that atopic patients normally resident in Arabia usually show skin sensitivity to extracts of pollen from *Chenopodium album*. In London grass pollen is the principal offender with concentrations of English grown *C. album* which is antigenically similar to the Arabian variety. In Kuwait on the other hand, *C. album* predominates and can therefore be expected to be a cause of symptoms, for grass pollen was only occasionally trapped.

As far as fungal spores are concerned, to a very large extent these are dependent upon the environmental temperature and will not grow above 29° C; thus the greatest concentration might be expected, if at all, in the cooler months. This was in fact the case, the most common fungal spores being those of the genus *Cladosporium* forming nearly 70% of the total spores counted; the highest concentrations occurred when the wind was from the north-west. As will be seen from Davies's chart, the month of November provides the highest average concentrations, when the temperature lies between 20—24° C and the mean relative humidity is between 55% and 65%. Davies asserts that the origin of the spore clouds in north-west of Kuwait is the more temperate regions where cereals and grasses are grown, but that the finding of clumps of spores containing large numbers of conidia also indicates production from a local source. These local sources could be twofold, from the experimental farms maintained by the Government and the other small, long-standing agricultural areas such as Jahra; but the probability is that the large grain elevators now operating next to Shuwaikh harbour are the more important source. Frankland and Davies feel that symptoms of hypersensitivity are precipitated when spore concentrations rise to 3,000 or more/m³: In Kuwait these concentrations have not yet been reached.

A further source of spore pollution of air is in the house dust, from which varieties of *Aspergillus* can be obtained. *A. niger* was the most common mould present when examined by Davies: The cooling effect of air-conditioning, by raising the relative humidity, will produce a dew-point, an excellent environment for the propagation of these organisms. Positive skin reactions to *A. niger* are not uncommon in allergic patients from Kuwait and neighbouring countries. While not yet considered a significant cause of hypersensitivity the summation effect of ambient temperature and humidity resulting from air-conditioning, creates a vaso-motor reaction in the nasal mucous membrane and a mild irritation. These effects are regarded by some, including the present writers, as of importance in the overall symptomatology of "allergic respiratory disease".

A balanced study of the clinical picture, short and long term pulmonary effects and the epidemiology of allergic conditions due to inhaled particles is still to be carried out. For the present the verdict on any one known allergen must still be "not proven". It has been repeatedly observed that Indo-Pakistanis living in Kuwait appear to be more susceptible, but they are known to develop heavy vegetation in their gardens as well as bringing with them a racial susceptibility which is commonly observed in the Indian sub-continent.

Before leaving the subject of air pollution it may be of interest and some value to speculate on the possible influence of *atmospheric ionisation*. With the current concern about the biological effects of non-ionising radiation particularly the high frequency electromagnetic waves and low intensity acoustic energy, it is worth reminding ourselves that all matter is made of energy-containing particles; the very air we breathe abounds with them and the space it occupies affords the means for their propagation. These forms of energy are necessary for normal biological processes but a distortion or substitution of the physical and chemical elements can play havoc with the delicate mechanisms governing plant and animal adaptability. This has been very well exemplified by the precautions against ionising and non-ionising radiation effects in space research. At the present time there are very great variations in the measurable quantities of atmospheric energy dependent upon altitude, humidity, and the pollution by man with the various forms of non-ionising radiation.

Certain areas of the world exhibit considerable changes in atmospheric ionisation, which occur during the passage of a "front", which is the junction of masses of air of differing temperature, pressure and humidity occurring at borders of a cyclonic depression. It has been suggested, to give but one example, that the often increased allergenic effects of moulds in a damp atmosphere are related to both the increased concentration and the heavy positive ionisation. Conversely, clear mountain air with a negative ionisation contains little allergenic matter and has long been regarded as beneficial.

With the widespread industrial, scientific and urban pollution of the air together with the effects of natural climatic changes and those artificially created indoors originally on a small scale but now over increasingly large areas (Houston astrodome), the civilised human being is bombarded with a variety of potentially stressful stimuli. These must account for some of the disorders of modern life ranging from frank neuroses to alterations in immunity mechanisms which cannot be related to purely infectious agents. Several generations of doctors have now been brought up to know the importance of the dynamic balancing forces which preserve the "milieu interieur" of Claude Bernard (1813—1878), but perhaps we are only just beginning to appreciate the invisible significance of the "milieu exterieur" particularly now that man is reaching beyond the atmosphere into space. The effects of all these stimuli are being felt to an increasing extent in the people of Kuwait, although some of them are not yet recognisable even if they are anticipated. There is still an opportunity for the planners to sit down with the producers of these stimuli to design effective means to substitute, reduce or annihilate them. "Those technical people who saw errors and kept silent during the past decade-and-a-half or those who, for certain reasons, lauded blindly instead of warning openly, have been the witting or unwitting allies of the forces promotive of serious technical and, therefore, economic problems [1]."

XVI. Conclusion

At the outset, Kuwait was considered as a small anomalous development in Eastern Arabia; in our present position it is apparent that several important principles emerge which have a bearing on a much wider range of situations throughout the world.

One of the most important points considered was the different meaning of urbanization in the developing world today when compared with the urbanization process recorded in nineteenth century Europe. Clearly Kuwait's affluence has affected the physical growth of the urban areas in the State, but there are contrasts with the West and parallels with the East which deserve elaboration. In much of south and south-east Asia, urban growth has involved the convergence of a variety of cultural and racial strains at the nodes of concentration. In many instances, the process of development has meant the intrusion of "foreigners" with a superior level of education and commercial acumen into the area, resulting in a degree of internal social stress. Whether those "foreigners" are Europeans, Chinese, Palestinians, or simply Christians seems unimportant: what is important is that a degree of dualism is established within the national economy, producing a "gradient of modernization" within the confines of a single national unit. In Kuwait the situation is as complex as in the million cities of south-east Asia for, as the Factor Analysis of Kuwait bore out, as well as the modern oil industry physically separated from the capital and major centre of population and employment, we have a further dimension in Kuwait represented by the two main types of immigrant reaching the state. Politically and economically the effects of these internal schisms are unknown, but our study of the ecology of life in Kuwait suggests important ways in which such a variety of races affects future planning. The range of diseases likely to occur in a population drawn from widely separated sources is simply one measure of the problem in Kuwait.

This form of development is not peculiar to Kuwait for lower down the Gulf we find exactly parallel situations emerging which look like following the pattern first laid out by Kuwait itself. Beyond the Middle East area, other small countries in course of rapid development are finding that many of Kuwait's problems are manifest in their situations. In this context, the cases of Hong Kong, and Singapore spring to mind together with many of the smaller Caribbean islands. Thus our study of Kuwait points to a second more general finding, that while an enormous influx of capital obviously facilitates physical development, a more rounded growth does not necessarily follow. *Dualism* can and will exist in the most affluent society unless proper accent is laid on education and manpower training.

It is a real fact that, if the public health and education of a developing nation is advanced beyond a certain threshold, the energy so generated becomes channelled almost inevitably into fierce political ambition unless alternative and acceptable outlets can be found. *The lesson of enlightened colonialism* tells us this; the benefits of assistance from agencies such as the World Health Organisation have led along the same paths. Unless sufficient industry of diverse character is made available together with the improvements in health and learning ability, intellectual and political explosion or massive emigration must ensue. We believe that this highlights the interdependence of both "have" and "have-not" nations, the need for regional planning of all elements of the individual countries' national development and inevitably the sharing of wealth to create opportunities for absorption of intellectual and physical man-power. Kuwait has shown the way to a limited extent, limited because of her own inadequate man-power pool and the juggernaut of rising internal costs once top heavy non-productive service industries, government or private sector, predominate. Added to this has been an understandable but unfortunate desire to emulate the spectacular architectural triumphs of their historical forbears and the more florid fancies of other western "enlightened" nations. One result of this policy is that within such a small country a great construction industry arises, but seeks only low-grade labour, which itself has to be imported, while relying upon non-Kuwait expertise for design and engineering. Once the industry is established, it progresses by its own dynamic momentum with the result that the establishment of worthwhile secondary industry is inhibited by the impulse to erect bigger and better structures which must be filled, not by industrially producing plants which require relatively less financial outlay and architectural expertise, but by increasing numbers of citizens supported by the relatively few producers.

We believe that this is the direction which Kuwait's development has tended to take so far, despite the warnings of its friends and advisers who have sought to guide the establishment of sound investments, a planned economy based upon diversification of industry and trade and education linked with technical and professional requirements. The same blind insistance on university status, the attainment of knowledge of the "humanities" rather than the techniques will create the explosive mixture so sadly seen elsewhere in the Middle and Near East to absorb a country's energy and wealth.

We have examined the paths which have led to what will eventually become a political rather than an ecological situation. The opening of world-wide communications, the creation of relative material wealth for everyman leading to an insatiable appetite; the support given by an effective system of health care, which has reduced the chance of death and disease to proportions almost comparable with European countries, all these have, in Kuwait, illustrated and exemplified the brittleness and waywardness of a development policy creating a new and wealthy nation in the "twinkling of an eye".

The ecology of a biological community must be stable enough for adaptation, if it is needed, to be acquired. The spectacular advance of science and materialism to-day hinders this capacity though seeming to transcend the need for it. We have shown how the structure of life in the new urbanized communities contrasts with the extraordinary tenacity and strength of those who were born in the desert. If Kuwait's resources of wealth and power are suddenly cut off, then there is a terrible risk that the nation and its parasites will die from desiccation before help could be brought. This fearful possibility should never be far from the minds of those who are responsible for the production of oil and gas and those who administer the state. Oil is an international political weapon but it should not be allowed to become one at a national level for, unlike the Suez Canal, the pipe-lines must remain open, regardless. There can be no return to the "good life" of the Badu; instead the ancient prejudices must mingle with new tolerances to preserve and improve the lot of man in the Middle East, to stimulate his productivity and self-reliance coupled with interdependence. We can take the analogy of the Republic of South Africa which desperately needs the African labour yet denies it human dignity: as long as Kuwaitis will not adjust to manual and semi-technical work, Kuwait will continue to rely upon non-indigenous labour. This dangerous reliance must give way to the assumption of the national's responsibility; this last will only come about in an atmosphere of challenge and consideration of many of the immigrants' problems, to allow them full citizenship. One challenge would be to stop looking outside Kuwait's borders for higher education, to grasp the need to train Kuwaitis at home to produce what is there for the harvesting, the oil and its products, the sea and underground water for irrigation, stockraising of sheep and goats, the Gulf for its fish and above all the design and construction of urban and suburban dwellings that do not demand exclusive reliance on external power for combatting the rigours or climate.

Such then is the cultural pattern of Kuwait, nearing the completion of the second decade of its existence as a modern nation. But what of the effects of such a culture upon the health of a youthful nation reft by tensions and schisms arising from national pride and political ambition, separating Arab from Arab? How is it possible to see behind the mask of both fortunate Kuwaitis and less privileged temporary residents, mindful of the history of other luckless immigrants, the slaves imported from Africa who have, even now, remained subtly excluded from the privileges given to those with the pure strain of Arab blood? These are questions which we cannot pretend to answer, yet which must govern the thinking and action of all who live under such conditions. Our main purpose has been to define the basic environmental factors, to relate the response of the individual to the constantly changing situations he meets and to emphasise those areas of public and personal responsibility which can be placed upon the appropriate shoulders to create a sense of purpose, which could catalyse a restoration of the qualities of Arabic culture which formerly led the world in intellectual and scientific progress. This is a desire that has been held by others as well as ourselves, among them no less an indefatigable optimist that the late Saba George Shiber who "... has constantly worked for a clearer and deeper comprehension of contemporary urban-architectural affairs in the Arab world" [1]. He was well aware of the interplay of the biological, engineering and sociological influences in developing natural and artefactual resources in arid desert countries, which had been neglected by the many who had brought development in Kuwait "to the point when urban suicide was at least incipient in the oil city" [1]. His untimely death in 1967 created a void which may not be filled. Nevertheless if his admirers can do one thing, it is to introduce into the learning material of all children in Kuwait an understanding of their ecology, how it can be preserved and adapted at a rate compatible with their ability. Changes are sweeping over Kuwait at astonishing speeds but it will be for Kuwaitis of all categories to study and assemble their own energies and resources. There is perhaps too much of a tendency to look upon this little country as a laboratory, for it to be dissected and rearranged as others would like; this is reflected already in large areas of housing development, hospital services and patterns of retail trade. Yet if it is to remain a viable nation something more is needed than just being the banker; a people which opposes integration will surely founder; assimilation of immigrants is essential for full development. Kuwait should not hesitate to bestow its citizenship on those proved worthy, until now a privilege restricted to those few who have waited an unconscionable time for it. Shiber, whose monumental "Kuwait Urbanization" [1] has irritated and enthralled according to one's persuasion, was conscious that he had left "many profound gaps which must be filled by the many specialists in the fields of human organisation, socialisation and urbanization". This we have, in a small way, attempted to do.

It is Maurice King [2] in Africa who has developed the theme of the "cross-cultural" outlook, which seems also to be a useful approach to a small unit like Kuwait. He points out that by the very nature of their education, undertaken away from their own people or in distant lands from those they later serve, doctors are not, in general, culturally adaptable to the community in which they work and it is therefore doubly important that besides their professional and technical expertise they learn to appreciate the traditions, history, culture and ambitions of the people they serve, so that they can communicate with them at all levels. This gap has been alluded to by the Egyptian psychiatrists who came to Kuwait in the nineteen-fifties and which made their task so difficult despite the common bond of the Arabic language. We as reporters of the Kuwait scene are aware of the barrier which separates us from the community of Kuwait despite attempts to understand the thoughts and actions of the many people with whom we have had contact and discussion, sometimes involving that peculiar empathy between physician and patient which goes farther than the mere appreciation of another's tongue, home and work environment. While King gives many examples from Africa of the need to see beyond "the visible parts of a strange culture", nowhere is this better exemplified than when trying to understand the existence of infant and childhood malnutrition in an apparently affluent society of Kuwaiti and immigrant Arabs in an oil company population. Whether they are traders, technicians or artisans, each brings to this community a different culture yet all of them, when enclosed within a new and dynamic organisation, are subjected to the stresses and temptations as well as having to respond to demands from less fortunate family members, often still in the land of their origin, for help and protection. It has been instructive to breakdown the income of a worker's family and to see the apparently satisfactory sums of money that he brings home eroded by the claims of the two cultures, the new one he has embraced and the one he left behind, both of which may have led him into debt. The pattern of spending has been a major factor in producing unevenness in the family's development and only very recently has attention been paid to this important field by the health education and other authorities. It should be recalled that in Kuwait the average per capita income is nearly 1000 dinars (£ 1000), that there is no income tax, that petrol is a little over £ 0.05 a gallon, and that electricity and water are now virtually free or at minimal cost. Yet despite this there are many families hard put to feed their children well, even if they know how. The loyalty of wider family groups often goes beyond the just and prior demands of the intimate members, and the small ones may suffer, to reveal their plight in the clinics and wards of the hospitals. That the authorities, both government and industrial, are now aware of this type of problem, is illustrated by the encouragement and welcome given to individuals and agencies to come to Kuwait to study the social and economic factors influencing ordinary people in their daily lives, their buying and selling, their houses, their migrations, and their needs. In the medical field the same can be said, but islands of malnutrition in a sea of apparent plenty may be like volcanic bubbles on the surface, containing underneath all the ingredients of clinical and symptomatic disease, which may burst into activity when disturbed by some apparently indirect irritant or precipitating unconnected illness.

The resulting emotional impact upon the public and indeed upon the medical profession is small compared to the dynamic advances of modern clinical and laboratory medicine and surgery which are taking place; thus the expenditure upon the latter is astronomical in comparison with the funds made available to study and educate a people to live a healthy and disease-free life in their homes. This is a problem we recognise in affluent societies everywhere whether it is alcohol, tobacco, drugs, traffic accidents, air and water pollution or psychoneurosis. But the dividends, if we may be permitted to use such an expression, can be incomparably greater, in terms of the current and future health and happiness of a nation, if the priorities are identified and supported.

We are well aware that the lessons arising from Kuwait's experiences are not applicable to many less fortunate countries, but perhaps they may serve to illustrate the folly of creating structures requiring unbalanced capital "investment" when the per capita expenditure on national health is only a fraction of that spent upon the resident of Kuwait. Figures for the total population suggest that in the 1967/68 budget for the Ministry of Health each resident was allocated 150 Dinars (£ 150 or $ 420) one of the highest allowances in the world. Yet much of this is in capital costs and salaries of expensively trained doctors and it would be salutary to estimate the return on the investment over a five year period since, for instance, the Al Sabah Hospital was opened in 1963. We suspect that the outlay could not be justified in economic terms but adhere to the view that these are not the only criteria by which expenditure on a nations's health should be judged. The massive upgrading in medical and health services, along with all the other paraphenalia of a sophisticated state must carry with it the desire of others to emulate; herein lies the danger, for the less fortunate of Kuwait's neighbours may not be in any position to do this. The result can be the draining away from these countries of what expertise they have; this can be painfully obvious when we study the medical services of these countries. Nevertheless, by themselves they would not be able to provide a challenging outlet to many of the young people who see their chance in Kuwait, where they hope they can achieve training either within the country itself or abroad once they have shown their potential value to the Kuwait authorities.

The difficulty must always lie in the gap between the "have" and "have-not" countries widening and such a practice may seem to do this. Yet in the long term a number of these people may one day return to their own country. By then they might be in a position to contribute to the later and slower development, and learning from their own experience, avoid some of the inevitable pitfalls that a faster and less discriminating expansion of social and health services inevitably incurs.

Given the means, the will and the ability to accept change, Kuwait has shown that no place on earth is uninhabitable in the sense that it can not support the ordinary ways of life. Not only can the desert support such a life, but it can be a remarkably healthy one; the guardians of Kuwait's development must ever be alert to identify and exclude those noxious elements that have become so much a part of our existence in the industrialised world.

References

Preface

1. Simmons, J. S., Whayne, T. F., Anderson, G. W., Horack, H. M.: Global Epidemiology: The Near and Middle East. Philadelphia 1954, pp. 357 (Kuwait pp. 267—278).

Chapter I: Urbanization and Population Growth in the Middle East

1. Hauser, P. M., in: The study of urbanization. Eds.: P. M. Hauser and L. F. Schnore. New York 1965, pp. 9—10.
2. Davis, K.: The urbanization of the human population. Scientific American 213 (Sept.) 40—53 (1965).
3. Turner, R. (Ed.): Indian's urban future. California University Press 1962, pp. 5—6.
4. U. N. — UNESCO: Urbanization in Asia and the Far East. Calcutta 1957, pp. 6—7.
5. Hawkes, J., Wooley, L.: History of mankind: Vol. 1 — Prehistory and the beginnings of civilization. London 1963, p. 414.
6. Gottman, J.: Megalopolis. London 1963.
7. Azeez, M. M.: Geographical aspects of rural migration from Amara Province, Iraq. Unpublished Ph. D. thesis, University of Durham 1968.
8. International Labour Office: Why labour leaves the land. Geneva 1960.
9. McGee, T. G.: The southeast Asian city. London 1967.
10. Meier, G. M., Baldwin, R. E.: Economic development. New York 1964, p. 326.
11. Boeke, J. H.: Economics and economic policy of dual societies. New York 1953, p. 4.
12. Jones, E.: Towns and cities. Oxford 1966.
13. Davis, K.: World urbanization 1950—1970: Vol. I — Basic data for cities, countries and regions. University of California. Population Monograph Series 4, 10—24 (1969).
14. — See Ref. [13].
15. Hartley, R. G.: Recent population changes in Libya: Economic relationships and geographic patterns. Unpublished Ph. D. thesis, University of Durham 1968.
16. Hauser, P. M.: See Ref. [1], pp. 1—2.
17. Berry, B. J. L.: City size distributions and economic development. Economic Development and Cultural Change 9, 573—588 (1961).
18. Fisher, W. B., in: Europa Publications, The Middle East and North Africa. London 1966.
19. Shiber, S. G.: The Kuwait urbanization. Kuwait 1964, pp. 19—24.

Chapter II: The Urbanization of Kuwait

1. Adams, McC. R.: The land behind Baghdad. Chicago 1965.
2. Dickson, H. R. P.: Kuwait and her neighbours. London 1956.
3. Dickson, V.: The wild flowers of Kuwait and Bahrain. London 1955.
4. Milton, D. I.: Geology of the Arabian Peninsula: Kuwait. U. S. Geological Survey Professional Paper 560 F. Washington 1963, pp. 3—4.
5. Trewartha, G. T.: The earth's problem climates. University of Wisconsin Press 1961.
6. Fisher, W. B.: The Middle East, Chapter 3. London 1961.
7. Trewartha, G. T.: See Ref. [5].
8. Banerji, B. N.: Meteorology of the Persian Gulf and Mekran. Calcutta 1931.
9. H. M. S. O.: Weather in the Mediterranean. London 1963.
10. Salem, S. M.: Neurological complications of heat stroke in Kuwait. Ann. trop. Med. Parasit. 60, No. 4, 393—399 (1966).
11. Ffrench, G. E.: Community health in Ahmadi. Kuwait Oil Company 1966, p. 15.
12. Milton, D. I.: See Ref. [4], pp. 4—5.
13. Statistical Abstract: Kuwait 1967, Table 88.
14. Ministry of Guidance and Information: Kuwait today. Nairobi 1963, p. 128.
15. Planning Board: Economic survey. Kuwait 1966, p. 15.
16. Statistical Abstract: Kuwait 1969, Table 95.
17. Planning Board: See Ref. [15], p. 17.
18. Statistical Abstract: Kuwait 1968, Table 71.
19. Lorimer, J. G.: Gazetteer of the Persian Gulf and eastern Arabia, Vol. I. Calcutta 1915, p. 2297.
20. Ministry of Guidance and Information: See Ref. [14], pp. 154—155.
21. Bowen-Jones, H. et al.: Survey of soils and agricultural potential in the Trucial States. University of Durham 1968, p. 137.
22. Statistical Abstract: Kuwait 1966, Table 94.
23. Lorimer, J. G.: See Ref. [19], p. 1074.
24. Planning Board: See Ref. [15], p. 17.
25. Blegvad, H., Loppenthin, B.: Fishes of the Iranian Gulf. Copenhagen 1944.
26. Lorimer, J. G.: Gazetteer of the Persian Gulf and eastern Arabia, Appendix C, Vol. I. Calcutta 1908.
27. Strong, C. S.: Pearl diving in the Persian Gulf. Travel 75, No. 5, 11—14 (1940).
28. Alexander, A. E.: Pearl fishing in the Persian Gulf. Gems and Gemology 6, No. 2, 38—41 (1948).
29. Bowen, R. le B.: Marine industries of eastern Arabia. Geographical Review 41, 384—394 (1951).
30. Berreby, J. J.: Le Golfe Persique. Bibliotheque Historique. Paris 1959.
31. Census of Establishments: Kuwait 1965, Table 94.
32. Statistical Abstract: Kuwait 1968, Table 98.

Chapter III: The Early History of Kuwait

1. Arrian: The Life of Alexander the Great. Penguin Classics 1958, p. 248.
2. Niebuhr, Carsten: Beschreibung von Arabien. Copenhagen 1772.
3. Niebuhr, C.: l. c.
4. Nasir I Khusraw: Safer Nameh. Cit. by Wilson, A. T.: The Persian Gulf. London 1959, pp. 87—90.
5. Ibn Batuta: Travels of Ibn Batuta. Cit. Wilson, A. T.: The Persian Gulf. London 1959, p. 91.
6. Wilson, A. T.: The Persian Gulf. Summary Chapters 7 and 8. London 1959.
7. Hakima, Ahmad, M. A.: History of Eastern Arabia. Beirut 1965, p. 39.
8. Lorimer, J. G.: Gazetteer of the Persian Gulf and eastern Arabia, Vol. I. Calcutta 1908, p. 1003.
9. Mahan, A. T.: The life of Nelson. London 1899, p. 12.
10. Donaldson, A. N.: A history of the postal service in Kuwait. Kuwait 1966, p. 19.
11. Brydges, H. J.: A brief history of the Wahauby. Vol. II. London 1834.
12. Newins, Ralph: The Golden Dream. London 1963, p. 90.
13. Stoqueler, J. H.: Fifteen months pilgrimage through untrodden tracts of Khuzistan and Persia, Chapter 3. London 1832.
14. Palgrave, W. G.: Narrative of a year's journey through central and Eastern Arabia. London 1868, p. 286.
15. Aitchison, C. U.: A collection of treaties, engagements and Sanads realting to India and Neighbouring Countries, Vol. II. 5th Ed. 1933, p. 262.

16. Lorimer, J. G.: Gazetteer of the Persian Gulf and eastern Arabia, Vol. II. Calcutta 1915, p. 164.
17. Dickson, H. R.: Kuwait and her neighbours. London 1956, p. 153.
18. Dickson, H. R.: Kuwait and her neighbours. London 1956, p. 277.
19. Freeth, Zahra: Kuwait was my home. London 1956, p. 33.
20. Fraser, L.: India under Curzon and after. London 1911, p. 102.
21. Dickson, H. R.: Kuwait and her neighbours. London 1956, pp. 450—451.

Chapter IV: The Economic Development of Kuwait

1. Duaij, A.: Pre-oil Kuwait. Undergraduate dissertation. Department of History. Keele University 1962.
2. Imperial Bank of Iran: Annual Report 1949.
3. Imperial Bank of Iran: Annual Report 1951.
4. Government Information Bulletin, 16 March 1968 (Arabic).
5. British Bank of the Middle East: Annual Report 1954.
6. British Bank of the Middle East: Annual Report 1955.
7. Rostow, W. W.: The stages of economic growth. Cambridge 1967, pp. 10—11.
8. Lorimer, J. G.: Gazetteer of the Persian Gulf and eastern Arabia, Vol. II. Calcutta 1915, pp. 1054—1055.
9. Statistical Abstract: Kuwait 1966, Table II.
10. Census of Establishments: Kuwait (Arabic) 1965, Table 1.
11. Planning Board: Economic survey. Kuwait 1964, p. 8.
12. I. B. R. D.: The economic development of Kuwait. Baltimore 1965.
13. Planning Board: Economic survey. Kuwait 1966, p. 13.

Chapter V: Population Growth in Kuwait

1. Lorimer, J. G.: Gazetteer of the Persian Gulf and eastern Arabia, Vol. I. Calcutta 1908, pp. 1051—1052.
2. Landen, R. G.: Oman since 1856. Princeton (N. J.) 1966, pp. 143—144.
3. Jordan: Census of Population, Vol. 1, Table 5.2, 1964.
4. Azeez, M. M.: Geographical aspects of migration from Amara Province, Iraq. Unpublished Ph. D. thesis. University of Durham 1968.
5. I. B. R. D.: The economic development of Kuwait. Baltimore 1965, p. 25.
6. Cipolla, C.: The economic history of world population. Chapter 4. London 1962.
7. Mountjoy, A. B.: Industrialization and under-developed countries. Chapter 2. London 1963.
8. Barclay, G. W.: Techniques of population analysis. Chapter 5. New York 1964.
9. Calverley, E. T.: The Arabian American Mission in Kuwait. Amer. med. Women's Ass., August (1950).
10. — See Ref. [9].
11. Dickson, H. R. P.: Kuwait and her neighbours. London 1956, p. 448.
12. Lorimer, J. G.: See Ref. [1], p. 1050.
13. — Gazetteer of the Persian Gulf and eastern Arabia, Vol. I. Calcutta 1915, pp. 2517—2554.
14. — See Ref. [13], pp. 2517—2530.
15. — See Ref. [13], pp. 2530—2539.
16. — See Ref. [13], p. 2554.
17. — See Ref. [13], p. 1052.
18. Kuwait Municipality: Annual Report 1966—1967, p. 162 (Arabic).
19. Statistical Abstract: Table 16, 1967.
20. Coale, A. J., Demeny, P.: Regional model life tables and stable populations. Princeton, N. J. 1966.
21. Ffrench, G. E.: Community health in Ahmadi. Kuwait Oil Company. Kuwait 1966, p. 16.
22. — Infantile gastro-enteritis in Ahmadi, Kuwait. Paper delivered at the 13th Regional Meeting of W. H. O. Kuwait 1963.
23. Shaker, Y.: Infantile gastro-enteritis in Kuwait. Paper delivered at the 13th Regional Meeting of W. H. O. Kuwait 1963.
24. Kanter, H.: Libya — a geomedical monograph. Berlin-Heidelberg-New York 1967.
25. Freedman, R., Adlakha, A. L.: Recent fertility declines in Hong Kong. Population Studies 22, No. 2, 181—198 (1968).

Chapter VI: The Ecology of Daily Life

1. U. N.: Industrial development in the Arab Countries. Symposium held in Kuwait, March 1966, p. 82, 1967.
2. Johnson, D. L.: The nature of nomadism. Department of Geography, University of Chicago. Research Paper No. 118, 1969.
3. Merner, P. G.: Das Nomadentum im nordwestlichen Afrika. Berliner Geographische Arbeiten 12. Stuttgart 1937.
4. Dickson, H. R. P.: The Arab of the desert. London 1949.
5. Dickson. V.: The wild flowers of Kuwait and Bahrain. London 1955.
6. Dickson, H. R. P.: Kuwait and her neighbours. London 1956.
7. — See Ref. [4].
8. Stocqueler, J. H.: Fifteen months pilgrimage through untrodden tracts of Khuzistan and Persia, Vol. 1. London 1832, p. 18.
9. Lorimer, J. G.: Gazetteer of the Persian Gulf and eastern Arabia, Vol. I. Calcutta 1908, p. 1050.
10. Shiber, S. G.: The Kuwait urbanization. Kuwait 1964, pp. 16—24.
11. — See Ref. [10], part 2, 1964
12. I. B. R. D.: The economic development of Kuwait. Baltimore 1965, pp. 88—89.
13. Shiber, S. G.: See Ref. [10], Chapter 7.
14. Azzam, O., Buchanan, C., Thysse, J. P.: Policies to be adopted for a master plan of Greater Kuwait. Report to the Municipality of Kuwait 1965, p. 10.
15. Thysse, J. P.: Reports of Dr. J. Thysse. Duplicated (Arabic) 1962, p. 10.
16. — See Ref. [15], p. 9.
17. Azzam, O., Buchanan, C., Thysse, J. P.: See Ref. [14], p. 5.
18. Lorimer, J. G.: See Ref. [9], p. 1074.
19. Berry, B. J. L. et al.: Urban population densities: structure and change. Geographical Review 53, 389—405 (1963).
20. Sjoberg, G.: The pre-industrial city — past and present. Glencoe, Ill. 1960.
21. — See Ref. [20], pp. 1—24.
22. Hauser, P. M., Schnore, L. F.: The study of urbanization. New York 1965, pp. 335—346.
23. Mabogunje, A. L.: Urbanisation in Nigeria. London 1968.
24. McGee, T. G.: The Southeast Asian city. London 1967.
25. Sjoberg, G.: see Ref. [20] and in: Hauser, P. M., Schnore, L. F.: See Ref. [22], pp. 213—263.
26. Fruchter, B.: Introduction to factor analysis. Princeton, N. J. 1954.
27. Moser, C. A., Scott, W.: British towns. Edinburgh 1961.
28. Russett, B. M., et al.: World handbook of political and social indicators. Yale 1967.
29. Klovan, J. E.: Q-mode factor analysis programme for small computers. Dept. of Geology, University of Calgary 1966.
30. Hartley, R. G.: Recent population changes in Libya. Unpublished Ph. D. thesis. Durham University 1968.
31. Jones, F. L.: Social area analysis: some theoretical and methodological comments illustrated with Australian data. Brit. J. Sociology 19, 424—444 (1968).
32. Klovan, J. E.: See Ref. [29].
33. — See Ref. [29].
34. Imbrie, J.: Factor and vector analysis programs for analysing geologic data. Office of Naval Research, Technical Report No. 6. Northwestern University 1963.
35. Berry, B. J. L. et al.: See Ref. [19].
36. Sjoberg, G.: See Ref. [20].
37. Breese, G.: Urbanization in newly developing countries. Englewood Cliffs, N. J. 1966.
38. Brush, J. E.: The morphology of Indian Cities. In: R. Turner (ed.) India's urban future. University of California 1962.
39. McGee, T. G.: See Ref. [24], p. 139.

Chapter VII: „Health and Disease"

1. Todd, Lord: Royal Commission on Medical Education. London H. M. S. O. 1968, pp. 404.
2. Siegfried, Andre: Germs and ideas. London 1965, pp. 98.

3. May, Jacques: The ecology of disease. New York 1958, pp. 327.
4. Pavlovsky, E. N.: Textbook of human parasitology. 5th edit. Moscow 1948.
5. Hoare, C. A.: Reservoir hosts and natural foci of human protozoal infections. Acta tropica 19, 281—317 (1962).
6. Maegraith, B.: Exotic disease. London 1965, pp. 361.
7. Jusatz, H. J.: Richtlinien für die Abfassung von Medizinischen Länderkunden. Arch. Hyg. Bakt. 147, 279—288 (1963).
8. — (editor): Geomedical monograph series. Vol 1: Kanter, H.: Libya. A geomedical monograph. Berlin-Heidelberg-New York 1967. — Vol. 2: Fischer, L.: Afghanistan. A geomedical monograph. Berlin-Heidelberg-New York 1968.
9. Hudson, E. H.: Non-venereal syphilis: A sociological and medical study of Bejel. Edinburgh 1958, pp. 203.
10. Gelfand, M.: The sick African. Capetown 1957, pp. 866.
11. Stamp, Dudley: Some aspects of medical geography. London 1964, pp. 103.
12. Guthrie, J., Forsyth, D. M., Montgomery, H.: Asiatic influenza in the Middle East. Lancet 2, 590—592 (1957).
13. McCreadie, D. W. A.: Asian influenza in Kuwait. Brit. med. J. 2, 684—685 (1957).
14. Kanaan, M. W., Wilson, P., Badran, M. S.: The Kuwait smallpox outbreak (1967). J. Kuwait med. Ass. 1, 210 to 228 (1967).
15. Abu-Sittah, S., Dahabrah, S.: Infant Kala-Azar in Kuwait. J. Kuwait med. Ass. 1, 118—122 (1967).
16. El-Gazzal, McCreadie, D. W. A.: Hydatid disease in Kuwait. Brit. med. J. 2, 232 (1962).
17. Mokhtar, N. A., Kavaquisor, L. K., Steinhart, L.: Cerebral hydatid cyst. J. Kuwait med. Ass. 3, 215—221 (1969).
18. Hudson, E. H.: Non-venereal syphilis: A sociological and medical survey of Bejel. Edinburgh 1958, pp. 203.
19. Salem, S. N., Shuibair, K. S.: Nonspecific ulcerative colitis in Beduin Arabs. Lancet 1, 473—475 (1967).
20. Mokhtar, N. A., Salem, S. N., Ramadan, J., Shakir, Y.: Left-sided amoebic liver abscess in a boy aged 18 months. J. Kuwait med. Ass. 1, 167—170 (1967).
21. Salem, S. N.: Clinical trial of oral dihydroemetine in intestinal amoebiasis. Trans. roy. Soc. trop. Med. Hyg. 61, 774—775 (1967).
22. — The irritable colon and its subgroup the redundant loop syndrome. J. Kuwait med. Ass. 1, 16—22 (1967).
23. Shaker, Y.: Preliminary study of infantile diarrhoea in Kuwait. Paper read at XIII Regional Meeting (Middle East) W. H. O.
24. Ffrench, G. E.: Infant diarrhoe in Ahmadi, 1963. Paper read at XIII Regional Meeting (Middle East) W. H. O.
25. Shaker, Y., Mohktar, N.: Subacute sclerosing pan-encephalitis in Kuwait. J. Kuwait med. Ass. 1, 64—66 (1967).
26. Hornabrook, R. W., Moir, D. J.: Kuru: Epidemiological trends. Lancet 2, 1175—1179 (1970).
27. El-Alfi, O. S.: Congenital malformation in Kuwait. J. Kuwait med. Ass. 2, 35—38 (1968).
28. Stevenson, A. C., Johnston, H. A., Stewart, P. M. I., Golding, D. R.: Congenital malformations: A report of a study of series of consecutive births in 24 centres. Bull. W. H. O. Suppl. Vol. 34, 1966.
29. Hamdan, Y.: Survey of blindness in Kuwait. J. Kuwait med. Ass. 2, 35—38 (1968).

Chapter VIII: Preventive Medicine in Kuwait

1.

1. New Hospital In Arabia. Lancet 1, 661—662 (1960).

2.

1. Kuwait. Statistical Abstract, 1968.
2. Abdel-Kader, M.: Prophylaxis of tetanus. Proceed. 5th Arab Medical Conference. Kuwait 1966.
3. Wilson, P.: The Kuwait small-pox outbreak: A preliminary report. J. Kuwait med. Ass. 1, 93—94 (1967).
4. Kanaan, M. W., Wilson, P., Badran, S.: The Kuwait small-pox outbreak (1967). J. Kuwait med. Ass. 1, 210—228 (1967).
5. — — — Cutaneous complications of mass vaccination against small-pox in Kuwait. J. Kuwait med. Ass. 1, 269 to 276 (1967).

6. Abu-Sittah, S., Dahabrah, S.: Infantile Kala-Azar in Kuwait. J. Kuwait med. Ass. 1, 118—122 (1967).
7. Thom, W. F. J. M.: Annual report of preventive medicine division. Kuwait Oil Company 1966.
8. Salem, S. N.: Clinical trial of dehydroemetine in intestinal amoebiasis. Trans. roy. Soc. trop. med. Hyg. 61, 774—775 (1967).
9. Walker, A. R. P., Koorhuf, H. J., Richardson, N. J., Hayden-Smith, S.: Letter in: Trans. roy. Soc. trop. Med. Hyg. 59, 483 (1965).
10. Forsyth, D. M.: Practical difficulties in the treatment of schistosomiasis in an Arab Community. Trans. roy. Soc. trop. Med. Hyg. 52, 439—445 (1958).
11. — Schistosomiasis as seen in Arabia and East Africa. East Afr. med. J. 40, 261—266 (1963).
12. Ffrench, G. E., Barnes, L. B.: Treatment of schistosomiasis with organic antimony Compounds. J. Kuwait med. Ass. 1, 288—294 (1967).
13. Kelain, Y. Z., Wilson, P.: Experiences with Ciba 32.644 B. A. (Ambilhar) in the treatment of schistosomiasis. J. Kuwait med. Ass. 2, 151—162 (1968).

3.

1. Selim, M. M.: School health services in Kuwait. J. Kuwait med. Ass. 2, 129—131 (1968).
2. — El-Shazli, M.: A survey of skin and venereal diseases in the school out-patient department. J. Kuwait med. Ass. 1, 67—74 (1967).

Chapter IX: Treatment Services

1.

1. Dickson, H. R. P.: The Arab of the desert. London 1959, pp. 664.
2. Freeth, Z.: Kuwait was my home. London 1958, pp. 164.
3. Dickson, V.: Wild flowers of Kuwait. London 1955, pp. 144.
4. Hitti, P.: The Arab. Princeton 1943, pp. 224.

2.

1. Wilson, A.: The Persian gulf. London 1929, pp. 285.
2. Elgood, C.: Cholera and plague in the gulf. Proc. Persian Gulf med. Soc. 43—45 (1955).
3. Ives, Edward: A voyage from England to India, 1753. London 1773.
4. Seetzen: Letters to Baron von Zack. Monatliche Correspondantz XI and XII July—Dec. 1805.
5. Kelly, J. B.: Britain and the Persian gulf, 1795—1880. Oxford 1968, pp. 911.
6. Donaldson, A. N.: History of the postal Service in Kuwait, 1775—1959. Kuwait 1966, pp. 19.
7. Onsi, A. A.: Medical services in Kuwait: Blood bank. J. Kuwait med. Ass. 3, 39 (1969).
8. — El-Alfi, O. S.: The ABO blood groups in Kuwait. J. Kuwait med. Ass. 2, 3—10 (1968).

Chapter X: Trauma, Temperance, Tuberculosis and Toxoplasmosis

1. Mylrea, S.: Kuwait before oil. Produced privately. U.S.A. 1952, pp. 161.
2. Thom, W. F. J. M.: Annual report of preventive medicine division. Kuwait Oil Company 1965.
3. Salem, S. N.: Alcohol beverages: Life destructive agents. Proc. 5th Arab. med. Conference, Kuwait 1966.
4. Dickson, H. R. P.: The Arab of the desert. London 1959, pp. 664.
5. Editorial: Kuwait health services for tuberculosis. J. Kuwait med. Ass. 1, 229—230 (1967).
6. Thom, W. F. J. M.: Annual report of preventive medicine division. Kuwait Oil Company 1966.
7. Booz, M. K., El-Gayar, A. R.: Tuberculosis of bone and joint in Kuwait. Proc. 5th Arab med. Conference, Kuwait 1966.

8. Ffrench, G. E., Curd, G.: A study of the ecology of toxoplasma gondi in Kuwait. J. Kuwait med. Ass. 3, 255 to 266 (1969).
9. — Toxoplasmosis in Ghana, II. W. Afr. med. J. 11, 171 to 197 (1962).

Chapter XI: Psychiatric Illness

1. Ministry of Guidance and Information: These are our hospitals. Kuwait 1963.
2. Kline, N.: Psychiatry in Kuwait. Brit. J. Psychiat. 109, 766—774 (1963).
3. Darwish, H.: Schizophrenia and social background. Proc. 5th Arab. med. Conference, Kuwait 1966.

Chapter XII: The Haemoglobinopathies

1. Lehmann, H., Hunstman, R. G.: Man's haemoglobins. Amsterdam 1966, pp. 331.
2. Gelpi, A. P.: Glucose-6-phosphate dehydrogenase deficiency in Saudi Arabia: A Survey. Proc. 3rd Middle East Paediatric Congress, Beirut 1963, p. 65.
3. Lehmann, H.: Personal Communication.
4. Moktar, N. A., Onsi, A. A.: Sickle-cell disease in Kuwait. J. Kuwait med. Ass. 1, 1 (1967).
5. Ffrench, G. E., Shalhoub, E. S.: Haemolytic disease in Kuwait. Proc. Persian Gulf med. Soc. 1964.
6. Hugh-Jones, K., Lehmann., H., McAlister, J.: Some experiences in managing sickle-cell anaemia. Brit. med. J. 2, 226—229 (1964).
7. Shalhoub, E. S.: Use of desferrioxamine B (desferal) in thalassaemia. Proc. Persian Gulf med. Soc. 1968.
8. Wyngaarden, J. B.: Regulation of cellular metabolism. Physiol. Pharmacol. Physians 1, 5 (1966).
9. Shalhoub, E. S., Ffrench, G. E.: Haemolytic anaemia due to glucose-6-phosphate dehydrogenase deficiency in association with viral hepatitis. Proc. Persian gulf med. Soc. 1966.
10. Shaker, Y., Onsi, A., Aziz, R.: The frequency of glucose-6-phosphate dehydrogenase deficiency in the newborns and adults of Kuwait. Amer. J. hum. Genet. 18, 609 to 613 (1966).
11. Gilles, H.: Personal Communication 1964.

Chapter XIII: Heat Illness and Desert Survival

1. Johnson, R. H., Corbett, J. L.: Survival secrets of the desert Bedouin. Geographical Magazine XLI, 540—546 (1969).
2. Edholm, O. G., in: The biology of human adaptability. Ed. by P. T. Baker and J. S. Weiner. Oxford 1966, pp. 541.
3. Shaker, Y.: Thirst fever with a characteristic temperature pattern in infants in Kuwait. Brit. med. J. 1, 586—588 (1966).
4. Chloremis, K., Danelatou, C., Maounis, F., Basti, B., Lapatsanis, P.: Paper chromatography for amino-acids in thirst fever. Helvet. paediat. Acta 14, 44 (1959).
5. Kempstone, Lieut. G. .B: Notes on a survey along the eastern shores of the Persian Gulf in 1828. J. Roy. Geog. Soc. V, 278—279 (1835).
6. Southey, Robert: Life of Nelson. London 1951, pp. 278.
7. Guthrie, J.: Aetiology of heat illness in Kuwait. Proc. 6th Internat. Congress Trop. Med. Malaria VI, 89—94 (1958).
8. Leithead, C., Guthrie, J., de la Place, S., Maegraith, B.: Incidence aetiology and prevention of heat illness on ships in the Persian gulf. Lancet 2, 109—115 (1958).
9. Salem, S. N.: Neurological complications of heat stroke in Kuwait. Ann. trop. Med. Parasit. 60, 393—399 (1966).

10. Schrier, R. W., Henderson, H. S., Tisher, G. C., Tannen, R. L.: Nephropathy associated with heat stress and exercise. Ann. int. Med. 67, 356—376 (1967).
11. Guthrie, J., McCracken, A. V.: A table for treating heat-stroke. Lancet 1, 682 (1960).
12. Leithead, C., Leithead, L. A..: Climatic effects in comfort, health and efficiency. A report to the Kuwait Oil Company from the Liverpool School of Tropical Medicine 1957.

Chapter XIV: Occupational Health

1.

1. Ezzat, A. M.: Occupational health in Kuwait. J. Kuwait med. Ass. 1, 300—311 (1967).

2.

1. Book of Genesis: Chapter 1, v. 14.
2. Book of Daniel: Chapter 4, v. 26.
3. Pliny the Elder (A. D. 23—79): Natural history.
4. Herodotus (B. C. 484—424): History. 1st English translation 1584.
5. Woolley, Sir Leonard: Ur of the Chaldees. 2nd ed. London 1950, pp. 210.
6. Longrigg, Stephen: Oil in the Middle East. London 1955, pp. 305.
7. Duckert, Geoffrey: In: Morgan, E. D. (Ed.): Early voyages and travels in Russia and Persia. Hakluyt Soc. 439—440 (1886).
8. Longhurst, H.: Adventure in oil. London 1959, pp. 286.
9. Anderson, J. R. L.: East of Suez. London 1969, pp. 288.
10. Daily Telegraph (London): March 26th, 1970.
11. Handschim, R.: Environmental aspects of skin cancer in employees. Aramco Report (Med. Dept.) No. 9156 (1963).
12. Medical Research Council: The carcinogenic action of mineral oils. Report No. SRS 306. H. M. S. O. 1968, pp. 251.
13. Dooley, A. E.: Lubricants in the food industry. The Med. Bull. (Esso) 27, 136—148 (1969).
14. Lijinsky, W., Epstein, S.: Nitrosamines. Nature 225, 21 (1970).
15. Swann, P. G.: Metal poisoning. Med. Bull. (Esso) 28, 144 to 157 (1970).

3.

1. Dickson, H. R. P.: The Arab of the desert. London 1959, pp. 664.

Chapter XV: Air Pollution

1. Truhaut, R.: Air pollution and public health. Triangle (Basle) 8, 109—116 (1967).
2. Lawther, P. J.: Air pollution in Kuwait. Assignment Report W. H. O. (Regional Office, Eastern Mediterranean) 1965.
3. Wilkinson, W. M.: Development of allergy in the desert. J. trop. Med. Hyg. 67, 16—18 (1964).
4. Davies, R. R.: Spore concentrations in the atmosphere at Ahmadi, a new town in Kuwait. J. gen. Microbiol. 55, 425—432 (1969).

Chapter XVI: Conclusion

1. Shiber, S. G.: The Kuwait urbanization. Kuwait 1964, pp. 643.
2. King, M.: The cross-cultural outlook in medicine. Medical Care in Developing Countries. Nairobi 1966, Chap. 4.

Illustrations

Fig. 1. Kuwait City in 1969 with the old harbour in the left foreground: immediately behind lies the Saif Palace of the Ruler. Note the high rise buildings in the city centre and the new corniche

Fig. 2. The port of Shuwaikh with the Government water distillation plant and grain elevators in the centre of the photograph

Fig. 3. The original British Political Agency building built at the beginning of this century. It is now the home of Mrs. Violet Dickson, C.B.E. a resident of Kuwait for 40 years and interpreter of the people, flora and fauna of Kuwait

108

Fig. 4. A Kuwaiti "Bum" or trading ship. Some of these are over 300 tons carrying capacity

Fig. 5. An earlier photograph of the City wall of Kuwait, hastily constructed by its citizens against the threat of the "Ikhwan" in 1920. Only a few of the original gates now remain (see Fig. 33)

Fig. 6. Part of the City in the 1930's. Note the space between the edge of the town and the wall. The historic value of this picture justifies its reproduction which is not of good quality

Fig. 7. The late Ruler of Kuwait, Shaikh Ahmad al Japir al Sabah, turning the silver valve to start loading the first cargo of crude oil. June 1946

Fig. 8. Bahrah No. 1 well, North Kuwait, 1936

Fig. 9. Retail stores and import/export offices, 1967

Fig. 10. The Kuwait Chemical and Fertiliser Company plant, Shuaiba industrial area, 1969 (foreground)

Fig. 11. Deep water port at Shuaiba

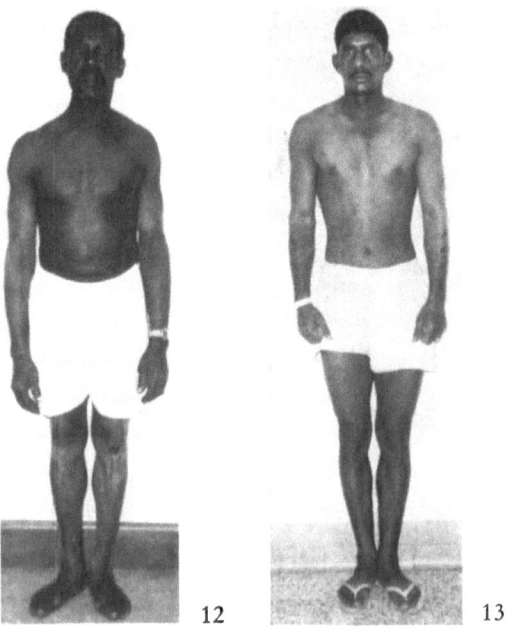

Fig. 14. A Kuwaiti Badu, winter clothing

Fig. 12. A Kuwaiti resident of negroid origin. A former pearl diver

Fig. 13. A Kuwaiti Badu. Note branding scars on his arms

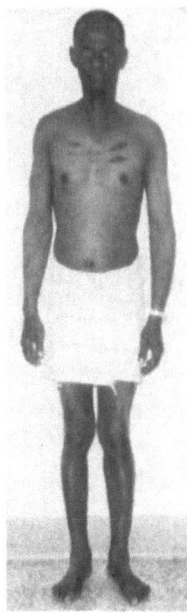

Fig. 15. A Kuwaiti resident of Omani origin. Note branding scars on his chest

Fig. 16. Traditional manner of raising water from large shallow wells in the small oasis town of Jahra

Fig. 17. Badu replenishing water supplies from trough provided by the Kuwait Oil Company at its pumping station at Abduliyah, forty km. west of Ahmadi

Fig. 18. Piped water supplies to a new housing area for recently urbanised nomads. Note the use of the traditional goat-skin water bags

Fig. 19. Preparing to move   Fig. 20. On the move

21

22

23

24

Fig. 21. Old houses in Kuwait before redevelopment

Fig. 22. A street in old Kuwait

Fig. 23. A stage in the development of Kuwait City

Fig. 24. 1951 view of Safat Square looking towards the Ruler's palace (Saif). The Suq runs between Safat Square and the sea

25

26

Figs. 25 and 26. The original bazaar or suq in Kuwait City. This still remains an important commercial and cultural centre

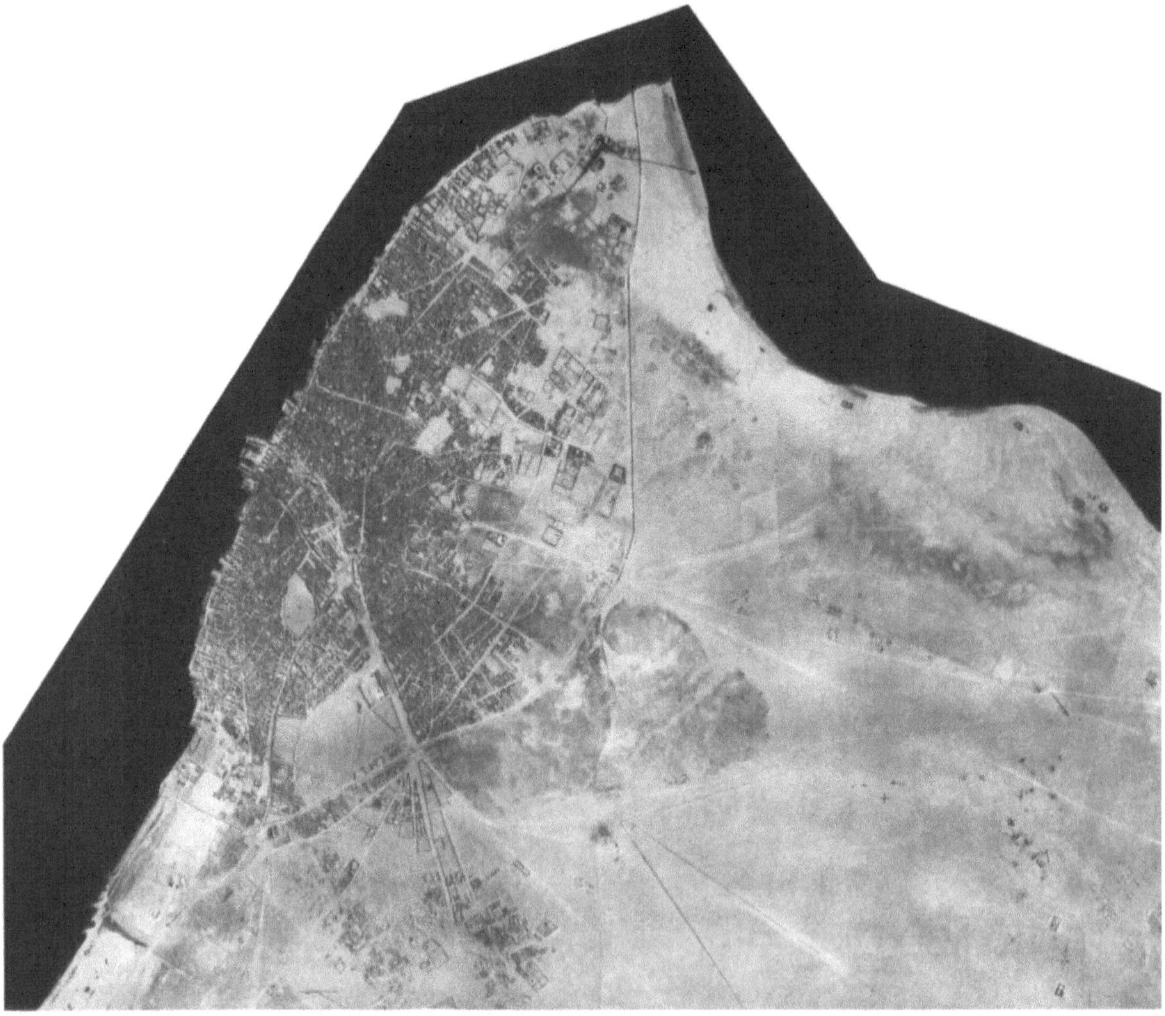

27

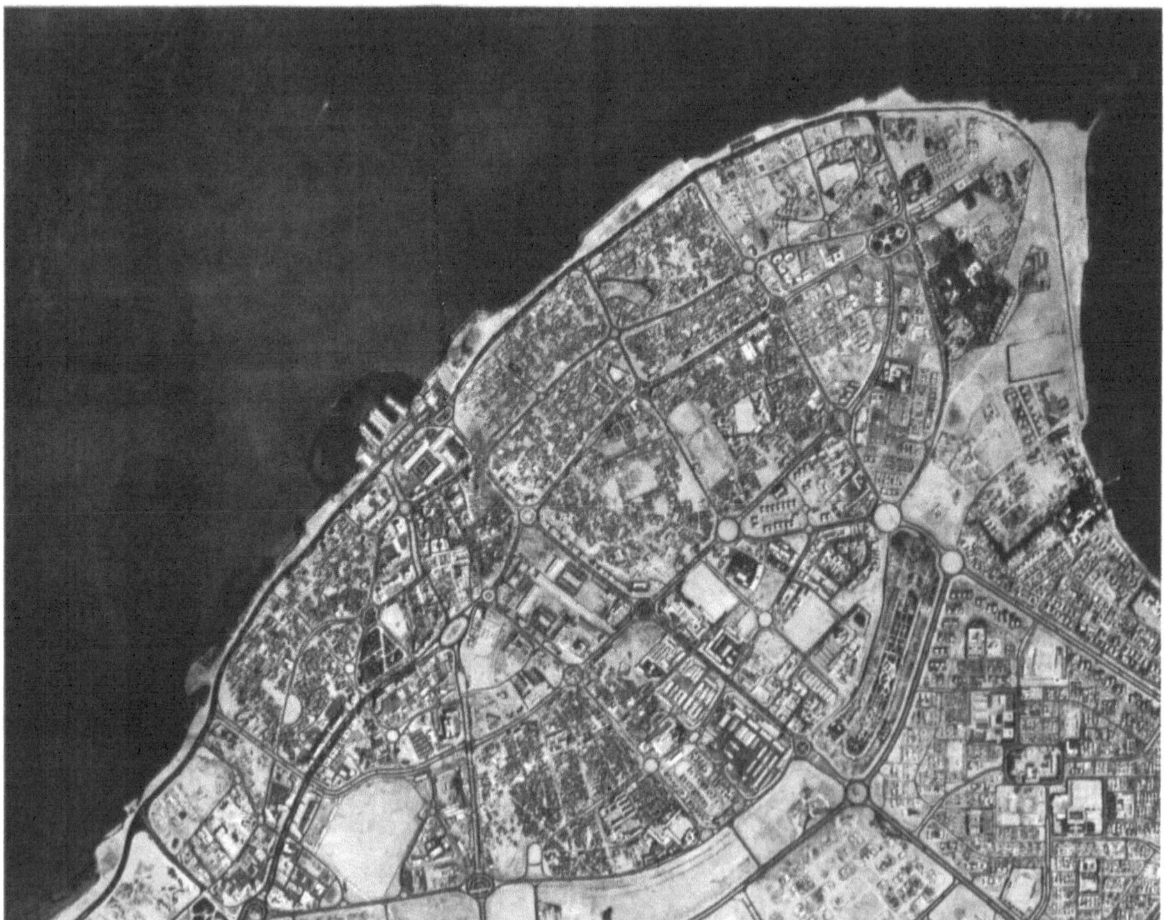

28

Fig. 29. The Saif Palace

Fig. 30. A view of the courtyard of an old Kuwaiti house taken in the mid-1950's. A few of the old houses remain

To page 114:

Fig. 27. The Old City of Kuwait in 1951. The large open spaces within the limits of the built-up areas are graveyards

Fig. 28. Parts of the 1951 Kuwait City remaining in 1967

8*

31

32

Figs. 31 and 32. The rapid evolution of the oil town of Ahmadi, over 20 years

Fig. 33. Fahad as-Salim Street, Kuwait City in 1969. In the foreground is the Jahra gate of the original city wall

34

35

36

Figs. 34, 35 and 36. Third Street, Ahmadi, 1951, 1957, and 1962

Fig. 37. The public gardens of Ahmadi

Fig. 38. A Kuwaiti Badu tent: the oil drum in the foreground was used to store water which was delivered by tanker-truck

Fig. 39. The remains of a sheep, killed for the Id-al-Fitr feast, thrown into a street in the town of Jahra in 1968

Figs. 40 and 41. Low income group housing development in Kuwait City

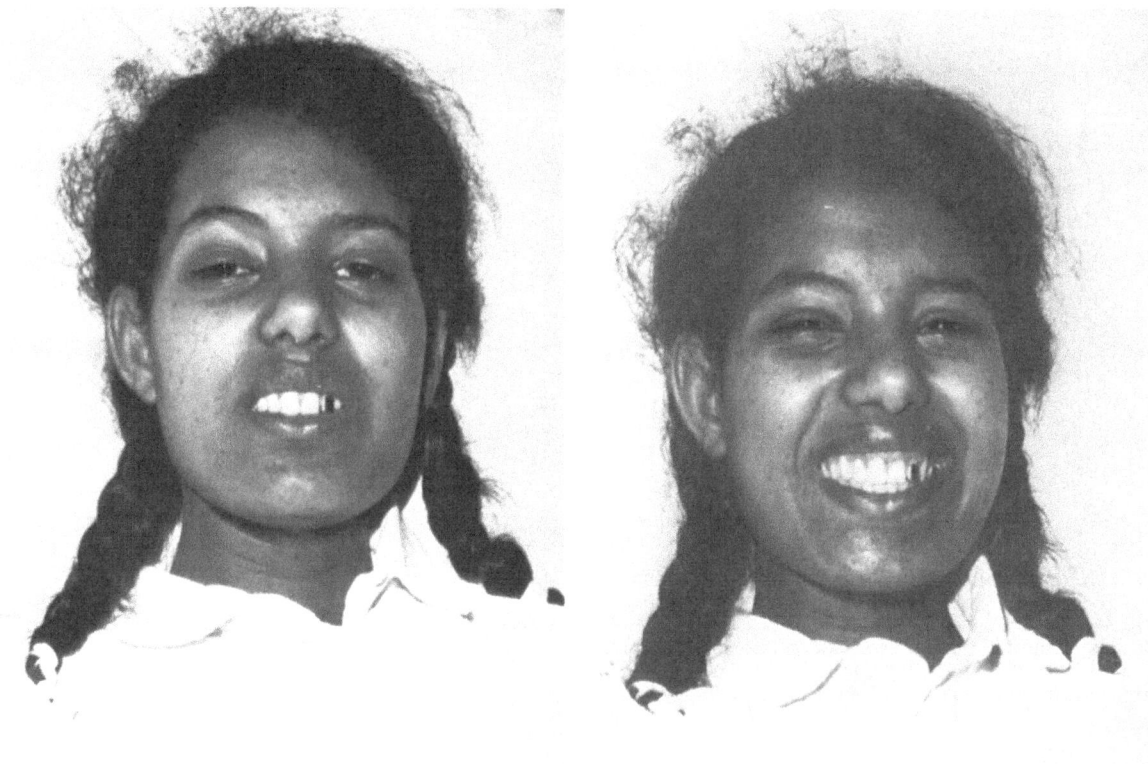

Figs. 42 and 43. An example of a comparatively rare type of muscle paralysis, Myasthenia gravis, probably due to an auto-immune process, occurring in an 18 year old girl. The photographs illustrate the effect of an intravenous injection of Tensilon (Edrophonium).
(Fig 42 before, Fig. 43 after injection)

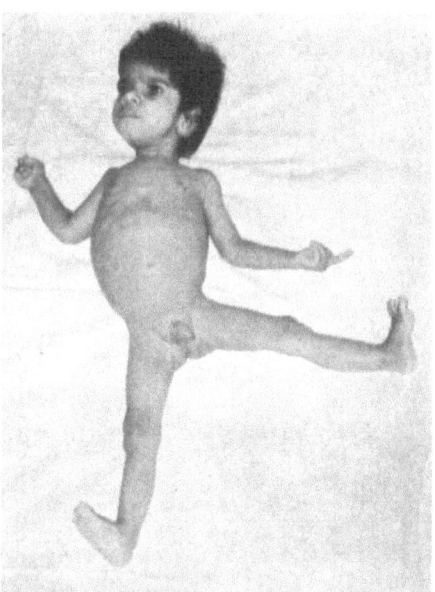 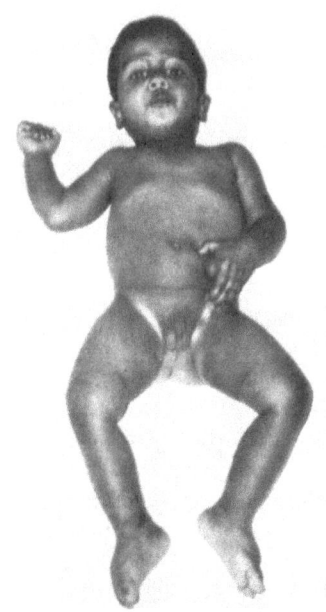

Fig. 44. Renal tubular disease in one of male Kuwaiti Badu twins

Fig. 45. His healthy twin

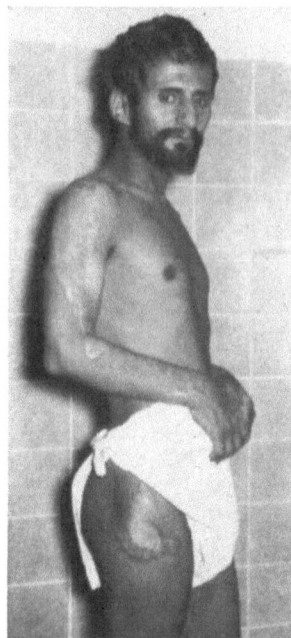

Fig. 46. The Southwell Hospital, Ahmadi; an example of a pavilion style, flexibly designed, fully air-conditioned, combined health centre and 250 bedded hospital, built in 1960. In the right margin is the helicopter landing pad. (Architects: Huckle and Durkin, London and Liverpool)

Fig. 47. Dimorphous leprosy in a Kuwaiti resident of Iraqi origin

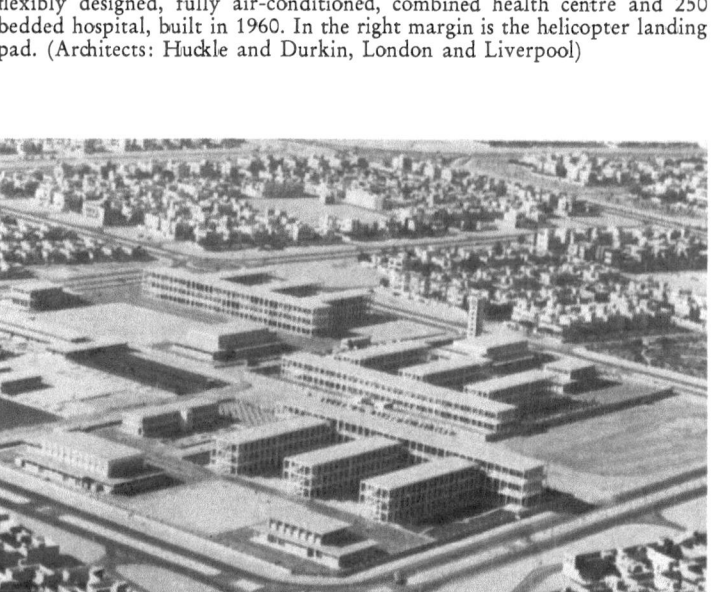

Fig. 48. Three examples of educational facilities in Kuwait City.
A primary school

Fig. 49. The State Secondary School, Shuwaikh

Fig. 50. A secondary school for girls

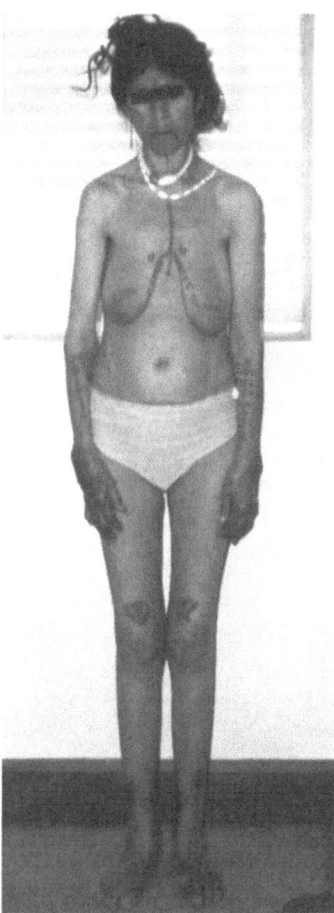

Fig. 51. Severe Rheumatoid arthritis and depression in a Badu woman. Note the tribal tatooing on face, neck, chest and limbs

Fig. 52. A fine example of a Kuwaiti member of an oil drilling crew

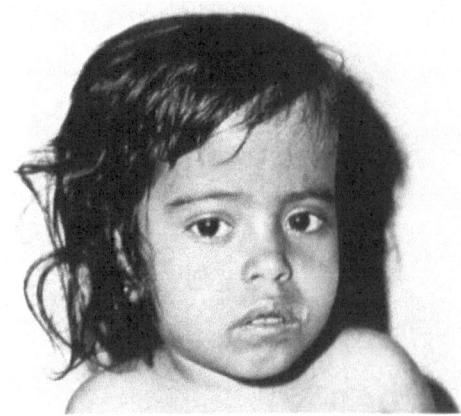
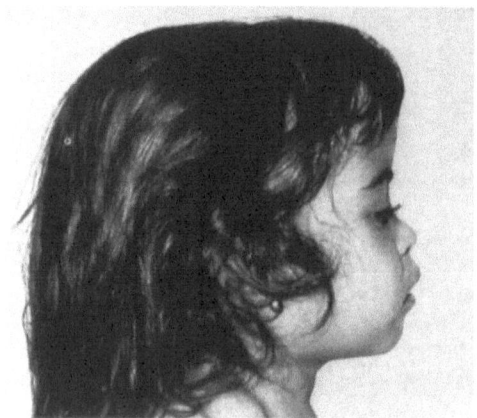

Fig. 53. Racial stigmata of β-thalassaemia major. S.M. female aged 3. Haemoglobin electrophoresis predominantly HbF. Singer test HbF = 75%

Fig. 54. Same patient as 53

Fig. 55. Fat-tailed Kuwaiti sheep

Fig. 56. The contrast of the shorn sheep, with little protection from sun radiation

Fig. 57. Check-list for desert driving at transport depot of the Kuwait Oil Company

Fig. 58. The Government petroleum refinery at Shuaiba adjacent to fertiliser plant (top right)

Fig. 59. An example of the generation of severe noise: Gas separators, not insulated

Fig. 60. Gas turbine pump of the 16 km undersea oil pipeline to the sea island terminal. It is housed in a special building which effectively contains the noise

Fig. 61. Exposure to sun radiation by an expatriate contrasted with the care taken by the local Arab workers

Acknowledgements

For permission to represent photographs, Figures 1, 2, 4, 5, 6, 9, 10, 11, 18, 21, 23, 24, 25, 26, 27, 28, 29, 30, 33, 40, 41, 48, 49, 50, 58, we are indebted to the Ministry of Guidance of the Government of Kuwait; and for Figures 7, 8, 16, 17, 34, 35, 36, 37, 46, 52, 59, 60, 61 to the Kuwait Oil Company, Limited.

Additional material from Kuwait,
ISBN 978-3-642-65174-8, is available at http://extras.springer.com